When Myth Meets Science

When Myth Meets Science

Legends That May Have Been Real

Dr. Aurelia Rowan

Mindful Pages

Published in 2025

ISBN: 9789371776417 (PB)
ISBN: 9789371779180 (eBook)

Published by

Mindful Pages
Imprint of Alpha Editions LLC
312 W. 2nd St #1834
Casper, WY 82601, USA

This book was developed using advanced writing tools to support the drafting and ideation process. All content has been thoughtfully curated, refined, and reviewed to ensure quality and relevance.

Content

Introduction
The World Behind the Veil

We like to imagine that we have broken free from the illusions that once bound humanity. Our satellites trace silent arcs through the night, algorithms chart the shape of our choices before we make them, and microscopes delve into the minute architectures of life. We tell ourselves we have outgrown myth—that those before us spun tales out of ignorance, that they trembled at thunder because they knew nothing of electricity, that they conjured gods merely to account for disease. From the lofty height of the present, we cast our gaze over the ruins of old beliefs and label them curiosities of a less enlightened age.
And yet—what if we have misunderstood entirely?

What if myth was never a crude draft of science, never a mistake to be corrected, but an entirely different instrument of knowing? Not a failed attempt at fact, but a language for meaning—one built of symbol, metaphor, and emotional resonance. Perhaps our ancestors did not confuse poetry for physics at all, but understood that some truths are not easily captured by measurement or proof. They may have perceived deep, recurring patterns in nature and human experience—truths too large or subtle for direct statement—and preserved them in the only form resilient enough to last: story.

The starting point of this book is a fracture in the modern mind's quiet certainties. We have been taught to think in binaries: myth is false, science is true; the past is fantasy, the present is fact. But this division, seductive in its neatness, is itself a modern myth—a narrative of progress that separates the analytical from the imaginative. It leaves little room for the possibility that myth and science might be complementary ways of apprehending reality. Science slices with precision; myth binds with association. Both seek to name the world, to make sense of our place within it.

We are surrounded by more information than any generation before us, yet many feel a gnawing absence—not a lack of data, but of meaning. We map the edges of the observable universe and decode the twisting code of life, but still struggle to situate ourselves within a larger, coherent story. This is not a failure of our tools; it is a loss of the mythic dimension. In turning away from myth, we may have cast aside one of humanity's oldest and most sophisticated frameworks for making sense of existence.

When Myth Meets Science does not propose that we take ancient tales as literal fact. Nor does it reduce them to nothing more than primitive psychology or decorative metaphor. Instead, it is an experiment in double vision—holding reverence in one eye and rigour in the other. It asks not what ancient people got wrong, but what they might have seen differently. It wonders whether certain myths preserve, in symbolic clothing, the record of real events, natural patterns, or states of consciousness that science is only now beginning to rediscover.

This is not a work of conspiracy or fantastical speculation; you will find no grand claims of extraterrestrial engineers or vanished continents. Rather, it practises a kind of disciplined wonder—an exploration that dwells in the space between scepticism and enchantment. Here the aim is not to declare final verdicts, but to listen: to the stories, to the data, to the recurring shapes and rhythms that appear across centuries, cultures, and minds. In that listening, new ways of seeing may quietly surface.

Think of this book as a spade for unearthing buried meanings. Each chapter is a dig site where fragments of story are lifted from the dust and examined under the light of modern discovery—be it in palaeontology, cosmology, neuroscience, or evolutionary theory. The goal is not to force one framework into the mould of the other, but to let each sharpen the other's outlines. Sometimes the alignments are uncanny, suggesting that memory—cultural or even biological—has travelled farther than we assume. Other times the differences are illuminating, revealing the limits of both ancient vision and modern analysis. In either case, the value lies in the dialogue, not the conclusion.

Your role here is active, not passive. You are asked to think like a detective of the unseen, to catch the faint threads linking disparate epochs and geographies. A myth may have wandered through countless retellings, reshaped by ritual or necessity, yet if it persists—if it echoes across worlds and eras—it invites the question: what enduring kernel allows it to survive?

Approaching this material is a little like cracking a code. You will be asked to read between disciplines, to inhabit the space between the likely and the possible. Dragons might emerge from the bones of long-extinct beasts; flood legends might whisper of real cataclysms; the duels of sky gods might reflect patterns in the heavens themselves. We will explore how myth might be both shaped by the human brain's appetite for pattern and, in turn, a shaper of how we think.

Myth, after all, is not simply a record of belief. It is a record of attention—what our forebears noticed, feared, longed for, and revered. It is a diary of how they placed themselves in relation to the cosmos, inscribed not in charts but in stories supple enough to cross oceans and centuries.

Science proceeds by breaking the world into parts. Myth, by contrast, gathers the scattered pieces into images. It does not ask for literal belief so much as a willingness to hear the symbolic undercurrent. And meaning—unlike measurement—belongs to the realm of analogy, ambiguity, and association: the native habitat of imagination.

We may imagine ourselves free of myth, but in truth we live inside new ones. The myth of perpetual growth. The myth of salvation through technology. The myth of progress as destiny. Like their ancient predecessors, these myths shape the architecture of our societies, though without the tempering humility of older traditions—those that recognised cycles, boundaries, and the unseen. Ours, more often, are myths of control.

The myths explored here are not replacements for science, nor relics to be sealed in glass. They are living artefacts with insights still sharp enough to cut. They do not tell us what to believe, but they remind us how to believe: with imagination yoked to inquiry, wonder paired with discipline. They urge us to stop pitting myth and science against each other and to see them instead as two hands shaping the same clay.

You will find no final answers here—only questions that stretch between strata of rock and strands of story, between the neuron and the nebula. Questions that suggest myth's purpose was never flight from reality, but orientation within it.

This is your invitation. Not to accept, but to attend. Not to dismiss, but to explore. To walk between worlds—not for certainty's comfort, but for coherence's challenge. Myth and science, each in their way, seek the real. And if we listen closely enough, we may hear where their voices meet.

Chapter 1
Memory in Disguise — What Myths Actually Are

We often consign myth to the realm of intellectual failure—a clumsy, pre-scientific attempt to grasp what lay beyond measurement. In our present imagination, myths are relics from an era without telescopes or thermometers, charming curios from minds that had not yet invented calculus or carbon dating. But this assumption says more about us than it does about those who came before. It betrays the extent to which we have pledged allegiance to a single way of knowing—empirical, linear, mechanical—and forgotten almost entirely the other modes of thought that once shaped human understanding.

What if myth was never a defective hypothesis, but an entirely different grammar? Not a falsehood born of ignorance, but a refined medium for transmitting layered meaning across centuries? Myth may not meet the modern standard of "fact," yet that does not make it untrue. It could be sacred fiction—truth rendered metaphorically, symbolically, spiritually. Here, we will consider myth not as a superstition to be dismissed, but as cultural code, carrying more than imagination: perhaps even fragments of deep, collective memory—memories of cosmos, Earth, body, and mind.

To take myth seriously, one must first learn how it works. Not what it declares, but how it speaks. As poetry obeys a logic different from prose, myth moves according to symbolic syntax. It trades in archetypes, not statistics. Its truth lies in the emotional, psychological, and relational—qualities that resist quantification. Claude Lévi-Strauss once proposed that myth is a language in its own right, built on its own structural laws. It conveys meaning not through chronological narrative or causation, but through pattern and juxtaposition. Myths rarely tell us what happened once; they reveal what happens repeatedly, across time and culture, as part of being human.

Carl Jung advanced this by framing myth as an expression of the collective unconscious—a symbolic reservoir shared across humanity. The hero's journey, the divine child, the cosmic tree—these are not whimsical inventions, but recurring motifs reflecting deep psychological realities. They emerge in dreams, in rites, in the carvings on temple walls and the dances of seasonal festivals. In this light, myth is not a fossil of error but a vessel of knowledge—a map not of the external world, but of the inner one.

Yet the inner and outer may not be so neatly separated. Myths might be more than inward symbols; they may function as compressed repositories of knowledge. Early humans, watching the heavens, marking seasonal change, tracking animal migrations, may have stored their observations not in almanacs but in allegories. The wrath of a sky god could encode the memory of a meteor strike. The serpent encircling the world might reflect celestial cycles. This is not pure conjecture—Australian Aboriginal traditions, for example, contain myths that map precisely onto geological formations created thousands of years ago, suggesting remarkable fidelity of memory over deep time.

Seen in this way, myth is less an archive of literal events and more a grammar of the "sacred lie." The phrase may sound contradictory—how can a lie be sacred?—but here the "lie" is not deceit; it is a vessel for truth that cannot be told directly. A sacred lie endures because it reveals rather than obscures, designed to be memorable, transmissible, and resonant on levels beyond the purely rational.

This challenges a modern prejudice—that empirical verification is the sole gateway to truth. The microscope cannot measure grief; likewise, the instruments of science may fail to capture the layers of meaning myth was made to hold. Yet, curiously, science's newest frontiers—in neuroscience, psychology, systems theory—are beginning to circle back toward patterns myth has long contained.

Consider the brain's compulsive search for pattern. We are not neutral recorders of reality; we are builders of narrative. We see faces in clouds, hear voices in static, spin causal chains from coincidence. This tendency, far from a flaw, likely offered an evolutionary edge: better to mistake the wind in the grass for a predator than the reverse. Pattern recognition was survival; storytelling was its cultural extension, embedding what to fear, what to seek, and how to navigate uncertainty.

From this angle, myth looks like the earliest survival technology. Before writing, before agriculture, humans told stories. Around them they wove rituals, laws, cosmologies. Myths carried warnings, strategies, moral maps—cognitive blueprints for thriving in a hazardous world. Unlike bare facts, stories cling; they travel, adapt, persist. In metaphor, they could encode ecological, astronomical, even medicinal knowledge, disguised yet enduring.

Altered states of consciousness likely played their part in myth's genesis. Across cultures, trance-inducing practices—through dance, fasting, drumming, plant medicine—were central. Such states produced visions: tunnels of light, animal guides, journeys through underworlds. Modern neuroscience may explain these as changes in brain chemistry or neural oscillations. But for ancient practitioners, they were revelations—messages from gods, glimpses beyond the veil. Whether viewed neurologically or mystically, their content—serpents, winged beings, death and rebirth—became the raw material of enduring myth.

If myth is a language, and the human brain is built to speak it, then it may follow rules—repeating structures that surface across cultures. This leads us to the idea of the cultural fractal: patterns recurring at multiple scales. A creation myth may echo the rhythm of seasonal festivals; a rite of passage may mirror a cosmology. Birth, death, renewal—the same symbolic logic appears in the life of the cosmos, the community, the individual. This self-similarity suggests that myths are less invented than emergent, arising wherever human minds encounter the vast and inexpressible.

And though the surface details differ—the gods' names, the monsters' shapes—the deep structure remains astonishingly constant. A hero descends and returns. The world ends and begins anew. A sacred mountain bridges heaven and Earth. These are not simply stories; they are organising frameworks, blueprints for perceiving reality's layered order.

Why does this matter? In an age drowning in information, myth offers something rarer: orientation. It reminds us that not all knowledge wears the badge of measurement, that meaning is no enemy of truth, that metaphor can point toward realities too intricate for reduction. One need not believe myths literally to learn from them; one need only recognise them as memory wearing the mask of story.

In the chapters ahead, we will consider myths that may encode the memory of genuine events—volcanic eruptions, celestial alignments, encounters with extinct creatures. But before that, we must understand their mechanics: why they take the shapes they do, how they embed themselves in the mind, why they endure. Myths are not only about dragons or floods or gods in the sky. They are about us—our fears, our longings, our grasping after the invisible. They are not inert tales from a vanished world. They are living codes. And if we learn to read them, they may still change the way we see.

The Grammar of the Sacred Lie

We moderns have an uneasy relationship with myth. On the one hand, we relegate it to the shelf of cultural antiquities—alongside bronze tools and burial masks—as a testament to humanity's intellectual infancy. On the other, we find ourselves compulsively drawn to it, embedding mythic structures into our films, branding, national narratives, and personal identities. Myth endures not because it reflects literal truth, but because it speaks in a language deeper than fact. It is not reportage; it is resonance. And to mistake it for failed history is to miss what it truly is: a metaphorical grammar—a symbolic system designed not to record what happened, but to reveal what *matters.*

The phrase *sacred lie* may seem jarring at first glance. It appears to pit truth against the sacred, fact against the divine. But this phrase, rightly understood, does not suggest deceit. It suggests a kind of higher fiction: stories deliberately constructed not for falsification or manipulation, but for preservation. What is preserved is not chronology, but cosmology—not the literal sequence of events, but the symbolic logic by which human beings make meaning. The sacred lie is a narrative crafted to survive time, trauma, and transformation. It is structured to speak across generations, often in compressed, encrypted form.

This is why ancient myths persist while contemporary headlines fade. Myths do not simply describe; they *encode.* They compress ecological observation, psychological insight, astronomical knowledge, and moral reasoning into structures that can be transmitted through oral tradition, ritual enactment, and symbolic storytelling. The form is not random—it is mnemonic, affective, durable. A myth, like a formula, carries more than its surface content. It functions as a vessel, shaped by a culture's needs, but often carrying cargo whose relevance outlasts its original context.

Claude Lévi-Strauss was among the first to articulate this with clarity. For him, myth was a system of oppositions—life and death, sky and earth, order and chaos—organised into stories that helped humans mediate the unresolvable tensions of existence. Myth, he argued, was not about belief, but about structure. The *content* of myths may vary widely across cultures, but their *form* often follows recurring logics. A flood purges the world so it can begin again. A hero journeys into the unknown and returns with knowledge. A divine figure sacrifices itself to redeem or recreate. These are not inventions of fantasy; they are patterns of human thought. They speak to the deep architecture of meaning.

Jung took a different, though complementary, approach. Where Lévi-Strauss saw myth as structural logic, Jung saw it as psychic expression. Myths, for Jung, were the dreams of cultures—archetypal stories that surfaced from the collective unconscious to give shape to inner experience. He recognised the strange familiarity of mythic figures: the trickster, the wise old man, the dark mother, the wounded king. These were not historical characters but symbolic types, representing energies and conflicts that repeat in the psyche. Jungian analysis did not reduce myth to neurology but acknowledged its symbolic precision. Myth, in his view, was how the soul narrates itself—how the unconscious speaks in images, not sentences.

Yet neither Jung nor Lévi-Strauss fully bridged the gap between these two views: myth as internal expression and myth as external system. To do so requires a more integrated approach—one that sees myth as both cognitive artefact and cultural record. Myths encode the *mind's architecture* while simultaneously preserving *observations about the world.* They are not simply metaphors for internal states; they are mnemonic maps of environmental, astronomical, and social knowledge. Indigenous mythologies around the world have been shown to track the migration of animals, the cycles of stars, the behaviour of weather systems. These myths were not speculative entertainment. They were operational knowledge rendered unforgettable through narrative and image.

This is why myths often feel simultaneously personal and cosmic. They touch on the inner experience of fear, hope, loss, and transformation—but they do so by attaching those feelings to stars, rivers, animals, and gods. The sacred lie operates as a symbolic interface: it connects the inner and outer, the known and the unknown. And its grammar—its underlying logic—is designed for transmission, not explanation. It does not argue; it performs. It does not persuade; it embeds.

Consider the flood myth. From Mesopotamia's *Epic of Gilgamesh* to the submersion tales of the Hopi and the warnings of the Māori, stories of overwhelming deluge pervade ancient cultures. Their structures vary, but they often share motifs: divine retribution, a chosen survivor, a vessel or refuge, and a rebirth of the world. These are not precise weather reports. But they may preserve real geological events—the sudden inundation of the Black Sea, meltwater pulses at the end of the Ice Age, tsunami-like consequences of cometary impacts. They also encode psychological responses: guilt, terror, renewal, survival. Myth, here, is not false history. It is symbolic compression—a story built to hold multiple layers of meaning simultaneously.

This ability to layer significance is what makes myth so enduring. It can speak to a child and a philosopher, to a tribe and a civilisation. Its grammar is recursive, like poetry or code. It uses metaphor not as decoration, but as syntax. A serpent may mean danger, wisdom, cyclical time, or cosmic energy. A mountain may mean axis, refuge, obstacle, or the spine of the world. Each symbol is polyvalent—capable of carrying different meanings in different contexts, yet always connected by thematic gravity.

That thematic gravity is not merely aesthetic. It has evolutionary utility. As organisms with highly developed pattern-recognition systems, human beings are predisposed to prefer coherence over chaos, even at the risk of false positives. We invent order. We impose story on randomness. What we call superstition is often just misfired pattern recognition—pareidolia in the face of complexity. But this tendency has survival value. It helps us anticipate threat, recall social norms, navigate uncertainty. Myth capitalises on this wiring. It structures information in emotionally charged forms—ritual, parable, cosmology—so that memory is enhanced and behaviour is shaped.

This brings us to a modern paradox. We live in an age that privileges data over story, but we remain creatures of story nonetheless. Scientific findings may be more accurate than myths, but they are also harder to remember, transmit, or act upon. A climate graph may show the rate of planetary warming, but a flood myth *feels* it. It connects the phenomenon to human stakes—punishment, cleansing, reckoning. This is not an argument for replacing science with story, but for recognising that they operate in different registers. One informs the intellect; the other anchors meaning. To communicate effectively, especially across cultural or ideological divides, we need both.

If myth is a kind of metaphorical grammar, then we must become better readers of that grammar. This means asking not "Did this happen?" but "What does this *structure* tell us about how people encoded their reality?" What did they fear? What did they observe? What did they revere? Myths may not be literal in the way modern science demands, but they may be *functional*—true in how they shape perception, behaviour, and cultural continuity.

We must also ask who has the authority to interpret myths, and for what purpose. Myths have often been domesticated—flattened into moral tales or nationalist propaganda. Yet in their original form, they were often subversive, ambiguous, and open-ended. They were not instructional manuals but dynamic performances. To read them properly requires a blend of critical literacy and symbolic fluency—an ability to navigate both the explicit and the implied. This is not the same as mysticism. It is a disciplined form of attention.

There are dangers, too, in interpreting myth solely through the lens of metaphor. Over-symbolisation can reduce real historical trauma to allegory, or sanitise violent myths into psychological fables. We must hold the tension between symbolic interpretation and ethical accountability. Some myths carry deeply problematic elements—misogyny, xenophobia, sacrificial violence. To treat these as merely poetic misses the fact that myths have material consequences. They shape norms, legitimise power, and define belonging. The sacred lie can be used to preserve wisdom—or to justify cruelty. It is not neutral.

The solution is not to dismiss myth, but to read it critically—neither literalist nor naïve. This involves understanding its function, recognising its syntax, and evaluating its applications. A myth can be simultaneously beautiful and dangerous, illuminating and deceptive. Its power lies in its grammar: its ability to make complexity feel comprehensible, to render the invisible visible. When we engage with myth on these terms, we begin to see that it is not a vestige of ignorance, but a resource for navigating uncertainty. It does not compete with science. It complements it. Where science tells us *how*, myth reminds us *why*.

Ultimately, the grammar of the sacred lie is not about deceit—it is about design. It is a symbolic system optimised for retention, resonance, and transmission. It teaches through indirection, conceals to reveal, simplifies to signify. Its durability lies in its structure, not its accuracy. And in this way, myth remains one of the most sophisticated tools human beings have ever developed for encoding—and decoding—reality.

From Shamanic Vision to Neuroscience

One of the most persistent misconceptions about so-called "primitive" cultures is that their visions of gods, spirits, and otherworldly realms were born of unscientific naivety. This view is both intellectually lazy and historically flawed. What modern neuroscience now confirms—often with great surprise—is that the vivid inner experiences historically associated with shamanism, prophecy, and sacred encounter are not delusions or fabrications, but deeply rooted in the neurology of the human brain. These experiences, far from being anomalies, emerge from identifiable cognitive and biochemical processes that have been part of the human operating system for tens of thousands of years. Ancient peoples did not mistake hallucination for reality; they mapped the contours of inner space with a fluency we are only beginning to understand.

At the centre of this inquiry is the shaman—not as a romanticised mystic or mere performer of tribal ritual, but as a sophisticated navigator of altered states of consciousness. The shamanic role existed across continents, from the Tungus-speaking peoples of Siberia (where the term "shaman" originates) to the ayahuasqueros of the Amazon, the San healers of southern Africa, and the oracular priests of ancient Greece. Despite profound cultural differences, the pattern is unmistakable: the individual who enters an altered state—

through trance, drumming, fasting, or plant ingestion—is tasked with accessing information, healing, or insight unavailable through ordinary perception. These states were not recreational. They were tools of cognition, rooted in communal need and neurochemical precision.

Modern neuroscience offers the language to understand what shamans once described in mythic terms. Functional MRI studies and EEG scans now show that altered states correlate with specific patterns in brain activity. One common signature is the suppression of the default mode network (DMN), a system of interconnected brain regions involved in self-referential thought, autobiographical memory, and internal narrative. When the DMN quiets, the boundaries of self begin to dissolve—a phenomenon often reported in mystical experience. This ego-dissolution is not metaphorical. It has physiological correlates and can be induced reliably through techniques ranging from deep meditation to the ingestion of substances like psilocybin, DMT, or ketamine.

These findings help explain why pre-modern peoples experienced their inner visions not as fantasy but as *contact*—with gods, ancestors, or other intelligences. The boundary between internal and external perception becomes porous in altered states. Visual hallucinations, synaesthetic experiences, and a sense of timeless unity are not signs of psychosis; they are part of what Roland Fischer once categorised as "ergotropic" states—high-arousal conditions that enhance symbolic thinking and pattern perception. From the inside, these experiences are indistinguishable from external reality. That distinction, so central to modern scientific thinking, is neurologically unstable in these states. The visionary landscape becomes a shared, usable realm—a place where stories, diagnoses, and cosmologies can be constructed with lasting cultural consequence.

The imagery reported across cultures in these states is strikingly consistent. Snakes, tunnels, spirals, beings of light, dismemberment and reassembly—these motifs repeat not because of cultural diffusion, but because of shared neurobiology. The visual cortex, when stimulated under certain conditions, produces geometric forms known as "form constants"—grids, spirals, and lattices—followed by complex iconic imagery. This layered experience—first abstract, then narrative—is documented in both Indigenous shamanic accounts and modern clinical reports from volunteers in psychedelic trials. What differs is the interpretive framework. Where a contemporary subject might describe "energy patterns" or "data fields," a shaman might speak of serpents or ancestral spirits. The underlying neurophysiology, however, is strikingly aligned.

The dismemberment motif in initiation rites, where the neophyte is symbolically torn apart and reassembled, is another case in point. Neuroscience understands this as a narrative translation of ego death and neural reorganisation. The subjective experience of fragmentation—feeling oneself dissolve, lose orientation, or become part of a larger consciousness—is mirrored in brain scans that show decreased connectivity in the self-referential regions and increased integration across previously siloed networks. Symbolically, this death-and-rebirth process becomes the core of many mythologies: the hero who descends, suffers, and returns transformed. But beneath the myth is a neurobiological pattern—one that has been ritualised across centuries into cultural grammar.

There is a growing body of research exploring how these experiences affect long-term behaviour and worldview. Clinical trials in psychedelic-assisted therapy have shown that a single deep experience of ego dissolution or cosmic unity can radically shift an individual's perspective on death, interconnectedness, and meaning. These outcomes mirror the traditional role of shamanic initiation, which was not merely a personal journey but a transformation into someone capable of holding contradictory realities: the world of spirits and the world of survival, the symbolic and the material, the private and the communal. These are not merely spiritual effects—they are reorganisations of perception that modern therapy now attempts to replicate.

What pre-modern cultures understood, perhaps more intuitively than we do, is that these altered states were *ontologically ambiguous*—neither wholly internal nor fully external. They required context, guidance, and integration. Ritual frameworks existed to structure the experience, provide symbolic interpretation, and prevent psychological fragmentation. Drumming rhythms, chants, mythic narratives, and communal validation all served to make these experiences intelligible. In contrast, the modern encounter with altered states is often context-less—induced recreationally, or in clinical isolation, without a larger mythos to absorb and transmute the insights. We have the instruments, but often lack the grammar.

This has led to a reevaluation of ancient epistemologies. Scholars such as Michael Winkelman, Iain McGilchrist, and Jeremy Narby have argued that shamanism represents an alternative cognitive mode—one that privileges holistic, intuitive, and symbolic processing over analytical dissection. Narby's controversial hypothesis, presented in *The Cosmic Serpent*, suggests that certain mythic images—such as intertwined serpents—might reflect encoded insights into molecular biology, specifically DNA. While this theory remains debated, it reflects a broader movement towards reappraising the epistemic validity of shamanic knowledge—not as science in disguise, but as a parallel mode of understanding, operating under different principles.

It is tempting, from a modern perspective, to reduce shamanic experience to neurotransmitters and neural circuits. But this reduction, while neurologically accurate, risks missing the point. What matters is not just what happens in the brain, but *what the experience means*, and how that meaning becomes culturally operational. The brain may produce visions; the culture decides whether those visions are madness, revelation, or initiation. In many traditional societies, these states were not pathological but prestigious. They marked the individual as someone who could mediate between dimensions, carry messages, or interpret omens. This social role gave structure to the chaos of altered perception and allowed the community to benefit from insights that might otherwise destabilise.

This reframing has practical implications today. As Western medicine begins to reintegrate altered states into clinical practice—through psychedelic therapy, trauma processing, and palliative care—it is rediscovering what Indigenous traditions never forgot: that consciousness is not a fixed state, but a spectrum; that healing often requires symbolic narrative; and that profound shifts in perception, when handled with care and ceremony, can catalyse deep transformation. The challenge is to develop new grammars for these old experiences—frameworks that bridge the neuroscientific and the mythological, the measurable and the meaningful.

We should be careful not to romanticise shamanism or its modern equivalents. Not all visionary experiences are profound, and not all insights gleaned in altered states are accurate or benign. There is always the risk of delusion, ego inflation, or cultural misappropriation. But neither should we dismiss these experiences as irrelevant or unscientific. The consistent presence of shamanic practices across human societies suggests an adaptive value—not just in healing or guidance, but in maintaining a sense of coherence between self, society, and cosmos.

Myth, in this context, becomes the narrative fossil of these visionary states. The flight of a shaman becomes the journey of a god. The serpentine vision becomes the cosmic dragon. The descent into the self becomes the underworld. These are not arbitrary stories. They are structured interpretations of real, embodied experiences—processed through ritual, encoded in symbol, and preserved in culture. As modern neuroscience continues to uncover the biological basis for these phenomena, it inadvertently validates what ancient traditions intuited long ago: that the mind is not bounded by waking reality, and that the stories born in altered states may carry insights deeper than logic can reach.

The Brain as a Pattern-Making Machine

The human brain is not a neutral recording device. It is not designed to see the world as it is, but as it *needs* to be seen for survival. Our species owes its evolutionary success less to strength or speed than to the ability to find meaning in noise—to extract coherence from chaos, signal from randomness. The brain is not a passive observer; it is a relentless pattern-making machine, wired to detect structure even where none objectively exists. This tendency is not a bug in our cognitive architecture. It is a feature—one that gave rise not only to language and science, but to myth itself.

This instinctive drive to identify patterns is what allows us to read a face, follow a trail, or anticipate the shifting dynamics of social alliances. But it also explains why we see shapes in clouds, hear voices in static, or infer intentions from random events. These phenomena, known as *pareidolia* and *apophenia*, are not trivial quirks of perception; they are evidence of a deep cognitive imperative to bind experience into narrative. The cost of a false positive—believing a rustle in the grass is a predator when it's only the wind—is negligible. The cost of a false negative—ignoring a real threat—is potentially fatal. Evolution has therefore tilted us towards meaning-detection, even at the risk of illusion. We would rather be mistaken than extinct.

What emerges from this biological bias is a cognitive environment where stories become not just entertainment, but instruments of survival. Narratives allow us to organise information temporally, causally, and morally. They provide frameworks for memory, for planning, and for social cohesion. A well-told story is a survival map—transmitting lessons, warnings, and emotional insights that allow a group to function more effectively. Myths are the distilled, time-tested versions of these stories. They are *patterns of patterns*, carried forward because they organise reality in ways that resonate across generations.

This helps explain why mythic motifs are so persistent. The archetypal hero's journey, the cosmic battle between light and dark, the wise old guide—these are not random inventions. They reflect recurring structures of experience, embedded in a species that has always had to make sense of uncertainty. The brain prefers familiar shapes: both visually and narratively. When confronted with ambiguous stimuli—whether in the environment or in life itself—it instinctively imposes order. The cost of cognitive dissonance is high; myths resolve it by stitching together cause, consequence, and meaning into a form that feels intuitively complete.

Modern neuroscience has made significant strides in identifying the neural substrates of this pattern-seeking behaviour. The temporal lobes, in particular, are involved in the integration of sensory data into meaningful wholes. Electrical stimulation of these regions has been shown to induce religious or visionary experiences, often involving the sense of a "presence" or the feeling of cosmic significance. This is not to say such experiences are false, but that they are neurologically scaffolded. The brain does not simply record what happens; it interprets. And it often interprets in ways that favour narrative, even at the expense of accuracy.

There is a critical distinction to be made here between information and *interpretation*. The brain receives data through the senses, but that data is not useful until it is interpreted—filtered through memory, expectation, and cultural framing. A thunderclap is not just noise; it becomes the voice of a god, a harbinger, or a symbol of wrath depending on the mythological structure imposed upon it. These interpretations are not arbitrary. They emerge from a feedback loop between biology and culture. Myths are, in this sense, the cultural crystallisation of neural tendencies. They are stories that match the shape of the mind.

This feedback loop is especially evident in the phenomenon of *confirmation bias*—our tendency to seek out evidence that supports existing beliefs while ignoring contradictions. In mythic thinking, this bias reinforces the cohesion of the story. Once a culture has established a mythic framework—say, that eclipses are omens or that certain animals are sacred—evidence is gathered to reinforce it. Anomalies are absorbed into the narrative, or reinterpreted to preserve its logic. This is not evidence of irrationality, but of cognitive efficiency. The brain prefers to preserve coherence. It uses myth to protect the structure of perception.

Yet pattern recognition, for all its evolutionary advantages, is not without risk. The same neural systems that allow us to detect real relationships can also generate spurious ones. This is especially apparent in conditions like schizophrenia or obsessive-compulsive disorder, where the brain's pattern-detection machinery runs in overdrive, producing delusions or compulsions. But even in neurotypical cognition, the propensity to connect dots—across time, space, and category—can lead to conspiracy theories, superstition, or dogma. Myths, at their best, stabilise perception; at their worst, they can calcify it.

The tension between myth and empirical reasoning is not a matter of true versus false. It is a question of *function*. Myth operates on a level that is affective, symbolic, and associative. It provides maps for meaning, not measurements of material reality. Science, by contrast, privileges falsifiability, replication, and specificity. But both arise from the same cognitive source: a pattern-seeking brain attempting to make the world intelligible. The distinction lies not in the impulse, but in the method.

Understanding the evolutionary origins of this impulse helps defuse the condescension often directed at myth. To say that myths are "just stories" is to miss the point. *All* human knowledge is mediated through stories—through metaphors, models, and analogies. Even scientific theories rely on narrative frameworks: particles behave like waves, evolution "selects" for traits, the universe "expands." These are stories too—compressed, conceptual metaphors that help us organise the otherwise ungraspable. The difference is that science updates its stories when new data arrives. Myth resists revision precisely because its value lies in continuity.

That continuity is adaptive in its own right. In highly volatile environments, stable narratives provide psychological anchoring. They offer meaning in the face of death, order in the midst of chaos. The myth of a just universe, for example, may not withstand philosophical scrutiny, but it enables moral behaviour. The belief in ancestral spirits may not be scientifically demonstrable, but it reinforces intergenerational accountability. These are not trivial functions. They are part of what makes societies cohere.

Yet it is important not to mistake adaptive value for epistemic validity. Just because a story promotes survival does not mean it reflects objective reality. The key is to recognise the distinction between *functional truth* and *factual truth*. Myths often operate in the former register. They are not accurate depictions of events, but useful guides for navigating life. Their persistence is less a testament to their correctness than to their *fit*—to the way they align with our cognitive preferences and social needs.

This brings us to the critical insight: the brain does not default to myth because it is irrational, but because it is *human*. Rationality is a layered capacity, not a starting point. Infants do not begin by testing hypotheses; they begin by forming attachment, recognising faces, responding to rhythm and tone. The brain is shaped by early experiences that are inherently narrative—who comforts, who threatens, who disappears and returns. These foundational patterns become the architecture upon which more complex cognition is built. Myth plugs into this architecture. It speaks the language of emotion, repetition, and archetype.

The modern world, with its deluge of data and fractured timelines, challenges this architecture. We are bombarded with information, but often starved of coherence. The result is a hunger for narrative—a vulnerability to simplistic myths, whether in the form of political ideologies, pop-culture archetypes, or pseudoscientific belief systems. These are not signs of regression. They are symptoms of a deeper tension between our evolved cognitive architecture and the epistemic demands of contemporary life.

What is needed, then, is not a rejection of myth, but a literacy in mythic thinking. We must learn to identify when we are operating within a narrative structure, and what that structure enables or obscures. This requires both humility and discernment. The brain may be wired for story, but we are not condemned to live by inherited scripts. We can interrogate the patterns we perceive, question the meanings we assign, and hold space for ambiguity without losing coherence.

In doing so, we reclaim myth not as dogma, but as dialogue—a dynamic interplay between brain, culture, and environment. The same circuitry that sees faces in clouds can also recognise symbolic resonance in poetry, ritual, and art. The same impulse that once feared thunder as divine anger can now marvel at its electrical precision. The pattern-making machine is not a flaw. It is the engine of our deepest insights. But it must be guided—by reason, by curiosity, and by a commitment to recognising when a useful story becomes a limiting one.

Myth endures not because it resists science, but because it speaks to a dimension science alone cannot satisfy: the felt sense that life means something, that events connect, that chaos contains order just beneath the surface. Whether that order is real or imagined is often beside the point. What matters is that we are pattern-makers by nature. And myth is the pattern that made us human.

The Cultural Fractal Theory

Step far enough back from any single myth and you begin to see a strange repetition. A flood washes away a sinful world. A trickster subverts divine order. A hero descends into darkness and returns with knowledge. Across continents and centuries, cultures that never contacted one another have generated structurally similar stories.

The details differ—names, landscapes, gods—but the underlying architectures are uncannily consistent. This recurrence is often explained away as coincidence, archetype, or the result of shared human psychology. But such answers, while not wrong, may be incomplete. What if myths do not merely resemble each other because they arise from similar minds, but because they *function* as fractals—self-similar patterns of meaning that scale across civilisations, transmitting encoded information across time?

This is the core of the Cultural Fractal Theory: the proposal that myths behave not just like stories or symbols, but like fractal systems—recursive cultural structures whose patterns repeat on multiple levels, from the personal to the cosmological, from tribal lore to state ideology. Just as fractals in mathematics exhibit self-similarity regardless of scale, mythic patterns may reflect deep structural symmetries that organise cultural memory. These patterns do not simply resemble each other aesthetically; they perform equivalent functions within different systems. They are adaptive, transmissible, and resilient. And they suggest the existence of what could be called an *ancestral software*: a set of encoded narrative algorithms that different cultures render in their own symbolic language, but which trace back to a shared cognitive template.

To explore this concept meaningfully, we must begin with the essence of a fractal: iteration that reveals continuity. Fractals are not mere repetitions; they are recursive designs where the macro-structure echoes the micro-structure. The branching of trees, the spiralling of galaxies, the geometry of lungs and river deltas—all demonstrate this principle. What emerges at one level is not discarded at the next; it is preserved and elaborated. Cultural myths, when seen through this lens, appear not as linear accounts, but as iterative codes. A creation myth may be mirrored in a rite of passage. A seasonal festival may retell the same structure as a cosmogony. A nation's founding legend may recapitulate the same grammar as a local folk tale. Each version scales the pattern up or down, but does not abandon it.

One compelling example is the motif of death and rebirth—a near-universal narrative skeleton. In Sumerian myth, Inanna descends into the underworld and returns transformed. In Egypt, Osiris is killed, dismembered, and resurrected as king of the dead. In Mesoamerica, Quetzalcoatl dies by fire and is reborn as the Morning Star. In Christian doctrine, Christ is crucified, buried, and rises on the third day. The pattern is not cultural ornamentation; it serves a recursive symbolic purpose. It encodes change—whether biological, seasonal, psychological, or societal. And it does so in ways that can be reproduced at multiple levels: the individual facing personal trauma, the community undergoing renewal, or the cosmos cycling through destruction and regeneration. The fractal logic holds.

The repetition of this motif is not evidence of historical borrowing alone. Many of these myths emerge independently, separated by geography and language. This suggests that certain structures recur not because they are taught, but because they are *inferred*—not imposed externally, but generated internally by minds wired to organise experience in recognisable ways. The brain seeks stability through pattern; cultures reinforce this search with narrative. The result is the spontaneous emergence of similar symbolic grammars—what Jung called archetypes, but which we might now frame as *fractal scripts* operating across cultural systems.

Importantly, Cultural Fractal Theory does not claim that myths are copies of each other. Rather, it posits that myths are expressions of recursive cultural functions, shaped by both environmental constraints and cognitive architecture. For example, flood myths—while often dismissed as simple weather stories—can be understood as fractal events encoding social reset. On one level, they may preserve geological memory of sea-level rise. On another, they symbolise moral cleansing, generational turnover, or the collapse of order into chaos before restoration. These levels are not exclusive. They scale together, reinforcing the same essential message: that destruction is the precondition for renewal.

Anthropologists have long noted the phenomenon of mythemes—elemental units of myth that recur across narratives. Claude Lévi-Strauss treated these as the building blocks of mythological logic, capable of being reconfigured but always preserving a binary structure (life/death, culture/nature, male/female). While valuable, this structuralist view can be enriched by fractal analysis. Where structuralism maps oppositions, fractal theory maps *scalability*—how those oppositions are re-expressed in different symbolic contexts, maintaining the same relational logic. It is not just that myth A and myth B share a binary; it is that the structure of myth A recurs in the social rituals, cosmologies, and self-conceptions of culture B. This recursive quality is what gives myth its coherence across scales and epochs.

Contemporary systems theory lends further credence to this model. In complex adaptive systems, local interactions give rise to global patterns, which in turn shape local behaviours. Cultures operate similarly: individual beliefs feed into shared stories, which reinforce institutional structures, which then influence personal behaviour. Myths are the symbolic infrastructure of this loop. They encode models of reality that are flexible enough to be locally interpreted, but stable enough to maintain transgenerational continuity. This balance between fluidity and form is the hallmark of a fractal system.

Consider the modern persistence of mythic structures in secular contexts. The hero's journey, formalised by Joseph Campbell and adopted wholesale by Hollywood, remains one of the most recognisable narrative templates in the contemporary world. Blockbuster films, political campaigns, entrepreneurial autobiographies—they all reproduce the same recursive arc: call to action, descent into challenge, transformative ordeal, triumphant return. The details change. The function remains. This is not cultural inertia. It is fractal adaptation—mythic grammar updated to modern syntax, but preserving its recursive logic.

The implications of this theory extend beyond myth studies. If cultural narratives are fractal in structure, then interventions at one level—say, through education, art, or ritual—can reverberate through other levels. This suggests that cultural transformation is not merely a matter of policy or ideology, but of *pattern recognition.* To shift a society's trajectory, one must introduce new patterns of meaning that are scalable, resonant, and structurally compatible with existing mythic frameworks. This is not manipulation; it is symbolic engineering—an ancient practice now ripe for conscious application.

There are risks, of course, in invoking "ancestral software" as a metaphor. It risks collapsing complexity into determinism, or treating myth as algorithm rather than as living story. But the point is not that myths are coded once and executed indefinitely. Rather, they are flexible scripts—recursive templates that cultures rewrite according to evolving needs, yet which retain a recognisable structure. The software metaphor holds only if we acknowledge its limitations. Myths are not programs to be run; they are grammars to be spoken, and reshaped through use.

Critics may argue that pattern recognition invites projection—that we see fractals because we want to. This concern is valid, but it misidentifies the issue. The point is not whether patterns exist in some objective sense, but whether they are *functionally operative*—whether they structure behaviour, memory, and belief in ways that can be tracked and analysed. Cultural fractals do not require mathematical precision to be real. They require only that they replicate meaning across scale. That replication, once observed, becomes a tool of both analysis and creativity.

The Cultural Fractal Theory reframes the study of myth from static taxonomy to dynamic systems thinking. It invites us to look not just at what myths *say*, but at how their patterns *scale*. It pushes us to ask why certain motifs persist—not just because they are emotionally satisfying or morally instructive, but because they reflect the recursive logic by which humans navigate the world. If this theory holds, it offers a framework for understanding how ideas survive, how civilisations encode their values, and how we might begin to interpret the persistent architectures of meaning that shape our inner and outer lives.

And perhaps most importantly, it proposes that behind the diversity of myth lies a shared symbolic geometry—a set of generative forms shaped not by chance or diffusion alone, but by the intrinsic dynamics of consciousness, memory, and culture. Recognising these forms does not diminish the richness of myth; it deepens it. It allows us to see each story not as an isolated artefact, but as a node in a vast, recursive network of meaning—an echo from the collective mind, still unfolding across time.

Chapter 2
The Dragon Code — Fire, Flight, and Fossils

The dragon remains one of the most enduring and contradictory inventions of the human imagination. It surfaces with uncanny frequency across the mythologies of nearly every ancient civilisation—whether in the imperial courts of China, the volcanic plateaus of Mesoamerica, the sagas of the Norse, or the oral traditions of West Africa. It may appear winged or limbless, serpentine or horned, dwelling in oceans or soaring through the skies, cast as either guardian or destroyer. Yet, despite the wild diversity of its forms, the dragon is instantly recognisable. This is no simple coincidence of myth-making, but a recurring cultural pattern demanding explanation.

What truly distinguishes the dragon is not merely its sweeping range of mythic appearances, but its astonishing persistence within the human mind. Unlike localised spirits or deities, it emerges independently across civilisations that had no contact, in environments and cosmologies utterly unlike each other's. How could such different peoples, scattered across centuries and continents, conjure creatures that never existed yet describe them with such precise similarities? What shared pressures—cognitive, environmental, or both—make the imagining of a dragon not only possible, but inevitable?

Here, the aim is not to romanticise dragons, nor to dismiss them as quaint relics, but to probe what their longevity reveals about human perception, memory, and survival instincts. Dragons may never have stalked the earth in flesh and blood, yet they live—vividly—in our minds, myths, and cultural codes. That persistence suggests they belong less to fantasy and more to the architecture of human awareness: a distilled response to danger, a narrative shape for the unexplainable, a symbolic embodiment of the boundary between safety and peril.

On first inspection, the dragon appears biologically absurd. It fuses traits that, in nature, seldom coexist: the coiled agility of a serpent, the aerial grace of birds, the scaled hide of reptiles, the intelligence we associate with mammals, and, impossibly, the capacity to breathe fire. But therein lies its strange power. As an amalgam, the dragon condenses into a single impossible predator every cue that humans are primed to fear—fangs, slit-pupilled eyes, fire, wings, the ability to strike from above. It is the apex threat distilled to its essence: more symbol than animal, more warning than wonder.

Neurobiology offers a clue to why such an image takes hold so easily. Human brains are exquisitely tuned to detect predators—an inheritance from an evolutionary past filled with snakes, raptors, big cats, and sudden environmental hazards. The amygdala, that small but potent hub of fear processing, reacts not only to real predators but to images and patterns that suggest them. The snake, especially, commands an outsized share of our attention. Experiments have shown primates spot snake-like shapes faster than almost any other visual cue—a legacy of survival in a world where venom could kill. Transposed into the realm of story, this deep neural reflex could easily give rise to a creature like the dragon: a composite predator, etched into memory by the same circuits that once kept our ancestors alive.

Yet dragons may owe their existence not only to psychology but also to encounters with the deep past. Before science named and classified fossils, bones were stories waiting to be told. In ancient China, vast reptilian remains unearthed from riverbanks or caves were taken as physical proof of dragons. We now know many of these belonged to dinosaurs—stegosaurus, protoceratops—but at the time they served as confirmation of beings already alive in the cultural imagination. Records of such finds go back at least to 300 BCE, with dragon lore evolving partly to make sense of these enigmatic relics. Ironically, the very science that later dispelled such beliefs grew, in part, from them.

This pattern is not confined to China. In the American Southwest, Lakota and Cheyenne traditions speak of "thunderbirds" and vast reptilian beasts—perhaps sparked by the fossil beds revealed by erosion. In medieval Europe, regions rich in fossil deposits often birthed legends of giants or cave-dwelling monsters. Ancient Greeks, stumbling upon massive bones in mountain caverns, imagined the remains of colossal guardians. At such intersections, the line between observation and invention blurs: the earth offers a fragment of its ancient life, and human culture gives it the skin and breath of a dragon.

Even the most improbable of draconic features—the fire-breath—may have roots in natural spectacle. While to modern minds a flaming reptile is an absurdity, to early communities watching volcanoes erupt, meteors blaze across the sky, or wildfires burst from lightning strikes, the image would have been all too plausible. In many early cosmologies, volcanism embodied divine wrath—fiery power emerging from mountain or sky, often personified in serpent or dragon form. Among the Mesoamericans, deities such as Xiuhcoatl wove together fire and serpent imagery, merging geological force with mythic predator. In this light, the dragon's flame becomes not an invention but an interpretation: a narrative casting of environmental trauma into the shape of a living threat.

Oral tradition played its part in fixing these images. Myths need not be true to endure; they need only be memorable, repeatable, and resonant. The dragon, uniting geological, biological, and spiritual danger into one emblem, meets all three conditions. It allowed communities to dramatise hazards, transform them into rites of passage, and enshrine them in art, ritual, and architecture. In many traditions, dragons stand at thresholds—guardians of treasure, keepers of sacred knowledge, sentinels between the wild and the ordered. To meet the dragon was to meet the liminal, the place where two worlds touched.

Yet dragons are not uniformly malevolent. In China, they are emblems of wisdom, power, and protection—masters of water and weather, patrons of fertility, allies of emperors. In Hindu cosmology, nagas embody a similar ambivalence, capable of both blessing and harm. This duality may reflect ecological realities: in flood-prone lands, serpentine beings symbolise water's life-giving as well as destructive potential; in arid or war-scarred regions, they may embody fire, conquest, or divine retribution. The core archetype remains; its meaning bends to the land and its people.

Crucially, the global spread of dragon myths does not demand a single point of origin. The evidence favours convergent imagination: the idea that humans, equipped with similar neurological wiring and shaped by comparable existential threats, independently produced myths with parallel structures. As in convergent evolution—where unrelated species develop similar traits to meet similar challenges—so too with myth: the dragon may be a shared invention born of shared human conditions.

Evolutionary psychologists such as David E. Jones and Lynne Isbell have argued that the dragon's composite nature—serpentine body, talons, wings, fire—bundles ancestral dangers into a single, unforgettable image. In fearing dragons, children learn by proxy to avoid snakes, cliffs, fire, and predatory animals. The dragon thus becomes a teaching device—efficient in memory, charged with emotion, and remarkably effective in survival terms.

Its usefulness does not stop at the physical. In countless stories, dragons are vanquished not by brute force but by cunning, virtue, or divine aid. They are externalisations of inner conflict, avatars of greed, fear, pride, or chaos. To slay the dragon is not to destroy the wild, but to master it—integrating it into a higher order. That is why so many dragons guard not mere treasure, but sacred spaces or hidden wisdom. They are ritual adversaries: obstacles placed in the path of transformation.

What emerges is no single origin tale but a many-layered account. The dragon is at once psychological and palaeontological, ecological and symbolic. It compresses into one enduring figure what early humans feared, encountered, and imagined—and what they most needed to remember. Its survival into the present suggests it is not a discarded relic of the imagination, but a vessel for cultural memory, a warning still worth heeding.

The study that follows will treat the dragon not as idle superstition but as a precise cultural instrument, shaped by evolutionary memory, ecological intelligence, and symbolic reasoning. Through case studies from China, the Near East, the Americas, and Northern Europe, it will trace the dragon's mutable form under differing environmental and social pressures. It will examine the brain's deep readiness to respond to draconic imagery, and explore how fire—whether punishment, purification, or sheer destructive force—magnifies its power. It will also consider the role of fossils as concrete yet misread artefacts that anchored the myth in physical reality.

To decode the dragon is not to strip away mystery, but to reveal the mind's remarkable capacity to metabolise the unknown. The dragon endures not as evidence of myth's naïveté, but of its accuracy—a symbolic early-warning system, forged from fear, shaped by landscape, and sustained by awe. That we still find its shadow in our stories and our dreams suggests it continues to serve us, whispering old lessons about danger, memory, and the strange, enduring logic of wonder.

Dragons Across Time and Culture

No creature in human mythology is more widespread, more protean, or more persistent than the dragon. It appears, with unnerving consistency, across civilisations that never met, across geographies that never touched, and across centuries divided by oceans of time. This is not a trivial observation. When an archetype manifests with such frequency and structural coherence—from Bronze Age Mesopotamia to the mountain temples of imperial China, from medieval Europe to the highlands of Mesoamerica—it demands more than casual dismissal. The dragon is not just a mythical beast; it is a cognitive artefact. And understanding its global recurrence requires a multidisciplinary investigation into how the human mind encodes fear, survival, and meaning.

The problem is not simply that dragons are common; it's that they appear where they should not. There is no historical evidence of cultural diffusion that could account for their shared iconography. There is no fossil record of winged, fire-breathing reptiles that might explain the similarities in material terms. Yet, again and again, the same core features recur: serpentine bodies, scaled skin, wings or fins, association with water or fire, territoriality, and immense, often supernatural power. This is not coincidence. Nor is it evidence of shared delusion. It suggests an underlying structure in how the human brain processes certain types of environmental and emotional information—particularly fear.

The Chinese dragon, or *long*, offers perhaps the most sophisticated and enduring representation. It is typically portrayed as a composite of existing animals: the body of a snake, the claws of an eagle, the horns of a deer, and the face of a camel or demon. Far from being a symbol of chaos, the Chinese dragon is an emblem of imperial order, seasonal change, and cosmological balance. It governs rainfall, river systems, and agricultural prosperity. For millennia, it has been central to Chinese statecraft, ritual, and art. Unlike its Western counterpart, it is not slain but revered—an auspicious force rather than an adversary.

Contrast this with the dragons of Europe: antagonistic, greedy, flame-spewing beasts to be vanquished by knights or saints. From Beowulf's nameless dragon to Fáfnir in Norse saga, from the serpentine forms of Greek *drákōn* to the hellish monster of Revelation, European dragons are overwhelmingly hostile. They are destroyers of order, hoarders of wealth, despoilers of land. Their defeat is framed not only as a physical victory, but as a moral and symbolic one—the triumph of civilisation over savagery, of Christendom over paganism, of virtue over vice. Yet, even in their villainy, these dragons follow an internal logic. They guard boundaries. They test heroes. They exist at thresholds—between forest and village, chaos and culture.

Mesoamerican dragon analogues provide yet another variation on the theme. Quetzalcoatl, the "feathered serpent" of Aztec and Toltec tradition, embodies wind, knowledge, and creation. He is both sky and earth, both divine and anthropomorphic. In the Mayan pantheon, the Vision Serpent acts as a channel between worlds, appearing to kings during rituals involving bloodletting or hallucinogens. Unlike the fire-breathing dragons of the West, these beings are intercessors—liminal facilitators of cosmological order.

They link the visible and invisible realms, legitimising royal power through spectral connection.

The Aboriginal Rainbow Serpent, one of the oldest known mythological beings on the planet, reflects a similar symbolic density. It creates rivers and landsçapes as it moves, brings rain, and punishes taboo violations with floods or drought. Though each version is culturally specific, the underlying motif—an enormous, serpentine being tied to water, creation, and moral balance—echoes across language groups throughout Australia. What is striking is the consistency with which these beings encode both environmental processes and social rules. They are not inventions for entertainment; they are functional myths, structuring life in practical and psychological ways.

How, then, do we explain the global recurrence of this archetype? The most tempting explanation is Jungian: that dragons emerge from the collective unconscious, from archetypal structures buried deep in human cognition. But while this accounts for symbolic consistency, it lacks specificity. A more productive approach begins with evolutionary psychology. Human beings evolved in environments where large predators posed existential threats. We were prey long before we were apex predators. The neurological systems that developed to detect and respond to danger—especially danger that was serpentine, fast-moving, or came from above—are still active. The amygdala, responsible for fear processing, is unusually responsive to snake-like shapes. Visual cortex studies show faster activation when subjects are exposed to serpentine curves compared to more neutral forms.

This has consequences. When a culture needs a symbolic representation of threat or chaos, it will often default to forms that trigger these neural alarms. The dragon may thus be a composite embodiment of evolutionarily embedded fears: snakes, large cats, birds of prey, fire, and natural disasters. It is not a fantasy per se; it is an *amalgamated memory*—an encoded warning signal made more powerful by the addition of supernatural scale. This would explain why dragons often appear in transitional myths—creation stories, flood myths, cosmological battles—where the stakes involve entire worlds, not just individual lives.

Yet biology is only part of the equation. Cultural ecology adds another dimension. The dragon's attributes—flight, fire, wisdom, vengeance—can be read as extrapolations of environmental phenomena. Volcanic eruptions, thunderstorms, locust swarms, floods, and even solar eclipses have all been interpreted in dragonic terms. In societies vulnerable to environmental instability, it is natural to project cosmic agency onto those forces. A dragon, then, is a personification of unpredictability—something vast, dangerous, and beyond human control. By turning it into a narrative figure, cultures make that danger manageable. They name it, contain it, ritualise it. The dragon becomes a grammar for the inexplicable.

Fossil evidence also plays a role. In parts of Central Asia and China, dinosaur fossils have long been accessible at the surface. Some early Chinese pharmacopoeias describe "dragon bones" used for medicine. These were, in fact, dinosaur fossils—grindable, marketable, and often embedded in ritual practice. In the American West, the Lakota are thought to have incorporated large fossil finds into their thunderbird lore. Such finds lend physical credibility to myth, embedding real-world data into symbolic interpretation. The dragon, then, becomes not only psychologically plausible, but materially reinforced. The presence of bones, regardless of their actual origin, becomes a referent for what must have once lived.

These diverse threads—neurological, ecological, archaeological—intertwine into a powerful explanatory framework. The dragon is not just a global myth; it is a multi-domain expression of how humans process danger, uncertainty, and anomaly. Its consistency across time and space arises not from diffusion or plagiarism, but from cognitive and environmental convergence. It is a culturally flexible yet structurally stable symbol—capable of being adapted to suit empire or tribe, creation or destruction, reverence or fear.

One might be tempted to argue that dragons are no longer culturally relevant, relegated to fantasy novels or children's films. But this misses the deeper point. The forms may have shifted—dragons now appear as kaiju, alien invaders, or genetically enhanced super-predators—but the structure remains. We still use dragon-like figures to process collective fear, from the climate crisis to geopolitical catastrophe. Our mythic vocabulary has not disappeared. It has adapted.

To understand dragons across cultures, then, is not to study superstition, but to examine the architecture of meaning. What we call a "dragon" is a projection surface—a mythic screen onto which cultures map their deepest concerns. It is less a creature than a cognitive scaffold, enabling societies to integrate threat, power, mystery, and transformation into a single, potent symbol. It speaks not only to what we imagine, but to how and why we imagine it. And it reveals, with startling clarity, the universality of the human impulse to name what terrifies us—and then to tell a story that makes it intelligible.

The Palaeontological Hypothesis

It is easy to assume that myth arises solely from imagination—projected fears, cultural metaphors, symbolic systems disconnected from material reality. But some myths may have emerged not in abstraction, but from observation—specifically, the observation of something inexplicable, ancient, and partially preserved: fossils. The palaeontological hypothesis proposes that ancient peoples encountered the fossilised remains of prehistoric creatures and, lacking a scientific framework to interpret them, incorporated these findings into myth. Rather than misapprehending reality, they mythologised it—assigning symbolic meaning to the bones of creatures that no longer existed but whose scale and strangeness defied mundane explanation.

This hypothesis does not attempt to reduce all dragon myths or giant narratives to fossils, nor does it imply a simplistic one-to-one correspondence between bone and story. It suggests, rather, that *some* myths may be responses to real encounters with the fossil record—culturally encoded interpretations of natural artefacts encountered in the course of mining, agriculture, or ritual excavation. In this view, myths are not fabricated fantasies but interpretative responses to tangible anomalies. And the fossil, stripped of its modern taxonomic context, becomes a mythic artefact: a symbol whose meaning must be inferred in a pre-scientific world.

The most frequently cited case study supporting this view comes from China. For centuries, fossilised bones unearthed in rural areas were known as *"dragon bones"* (龙骨, *lónggǔ*) and were ground into powder for use in traditional medicine. These bones, often from dinosaurs of the Cretaceous and Jurassic periods, were believed to carry healing energy. Local apothecaries and farmers collected them from riverbanks and eroded hillsides, particularly in the provinces of Henan, Gansu, and Sichuan—regions now recognised for their rich dinosaur fossil beds. Some bones were as large as femurs from a *Mamenchisaurus* or jawbones from a *Psittacosaurus*—fossils that, viewed without palaeontological context, resemble nothing living. The idea that these belonged to dragons—revered beings in Chinese mythology associated with rivers, mountains, and the cosmic order—was not illogical; it was an act of pattern recognition using the symbolic vocabulary available.

More than anecdotal, this tradition was institutionalised. Chinese pharmacological texts dating back to the Han dynasty reference "dragon teeth" and "dragon bones" as ingredients. Some tombs contained fossilised remains deliberately included in burial rites, suggesting that the bones were seen not merely as curiosities or medicines, but as ritual objects. It is not unreasonable to propose that such material discoveries reinforced and possibly shaped the visual and symbolic conventions of Chinese dragon imagery: elongated bodies, ridged spines, clawed feet. While dragons were already present in Neolithic pottery and jade carvings, their evolving anatomical complexity may reflect cumulative fossil encounters interpreted through artistic traditions.

In the American West, a different but equally compelling narrative emerges. Indigenous peoples such as the Lakota, Cheyenne, and Blackfeet have oral traditions featuring monstrous beasts—gigantic birds or serpents, often with talons and reptilian characteristics—that lived in the deep past and whose remains could still be found in the land. These stories are often dismissed as metaphorical or cosmological, but their geographic correlation with major fossil sites invites closer scrutiny. The badlands of Montana, Wyoming, and South Dakota—regions central to the oral histories of these nations—are among the richest fossil fields in the world, containing specimens from *Tyrannosaurus rex*, *Triceratops*, and other large Cretaceous animals. It is entirely plausible that weathering and erosion periodically exposed bones that would, to any observer, resemble the remains of gargantuan, long-dead beasts.

Adrienne Mayor, a classical folklorist and historian of science, advanced this line of inquiry in her pioneering work *The First Fossil Hunters*. She argues that many classical accounts of griffins, giants, and dragon-like beasts in the Greco-Roman world were based on fossil finds. She cites numerous examples in which ancient Greek and Roman texts describe skeletal remains discovered in caves or mountain passes, often interpreted as the bodies of heroes, monsters, or gods. Pliny the Elder, for instance, describes a dragon killed in India that was 120 feet long—details that, while fanciful in tone, suggest some basis in observation, perhaps even of *sauropod* bones. Mayor also documents how Scythian gold miners in Central Asia reported the existence of hybrid creatures—part lion, part eagle, with bony crests and sharp claws—that align strikingly with

Protoceratops fossils found in the same region. These fossils, partially exposed in sedimentary layers, lie near historical gold-mining sites, supporting the hypothesis that early encounters with palaeontological remains informed local myth.

What makes the palaeontological hypothesis compelling is not that it seeks to explain myth away, but that it restores material context to symbolic systems. Myths are not created in a vacuum. They are shaped by the terrain, by visible anomalies, by shared perceptions of disruption or mystery. A fossil, especially one only partially exposed or fragmented, becomes a cognitive riddle. It forces an interpretation. If it resembles a vast claw, a horned skull, a ribcage the size of a man—it demands narrative integration. And in pre-literate or non-scientific societies, narrative integration means myth. The fossil becomes evidence for what the culture already senses: that the world was once different, that powerful forces once roamed the earth, and that their remnants remain—both as caution and as wonder.

It is critical, however, not to overextend the model. Not every dragon or monster myth must trace its origins to fossil encounters. Many clearly do not. Myth can arise from dreams, trance states, allegorical structures, or environmental metaphor. The role of the palaeontological hypothesis is not to replace these explanations but to supplement them. It reminds us that the line between natural history and cultural memory is often porous. Just as contemporary science seeks to interpret fossils within geological and biological frameworks, ancient societies interpreted them within cosmological and moral ones. The object remains; the meaning shifts.

This opens a broader methodological question: how should we read myth in light of palaeontological data? It requires a shift in hermeneutics—from reading myth as symbolic abstraction to reading it as material response. When a culture speaks of an "earth serpent" that shakes the ground, it may be referencing tectonic activity. But if it speaks of bones from a great beast unearthed by a storm, it may be referencing a fossil. The ability to distinguish between these layers of meaning—to ask not only *what a myth means*, but *what it responds to*—is essential for understanding the deep memory structures embedded in human storytelling.

There is also a philosophical insight here worth underlining. The palaeontological hypothesis suggests that even without microscopes or taxonomies, humans are capable of recognising and preserving deep-time realities through story. Myth, far from being an obstacle to science, may be its precursor—a scaffold for interpreting the ancient unknown in a way that integrates memory, environment, and meaning. In this light, myths about dragons and giants are not epistemological failures, but early attempts at knowledge: not just tales told to entertain or explain, but mechanisms for encoding observation, awe, and speculation about the world before humans.

As modern science continues to reconstruct Earth's prehistoric past, it does so with tools that formalise what mythic cultures approached symbolically. Today we name species with Latin binomials; then, they named them Thunderbirds or Dragon Kings. But both systems are responses to the same phenomenon: evidence that the earth was once home to creatures utterly unlike ourselves, and that their remnants remain as artefacts of deep time. The dragon, in this context, becomes a fossil of thought—an expression not of error, but of interpretation, shaped by a human need to make sense of what the world leaves behind.

When we consider that some of our most enduring mythological figures may have begun with the discovery of a strange bone protruding from the ground, we are reminded that myth and science are not antithetical. They are parallel modes of insight, grounded in the same cognitive impulse: to find meaning in what is buried, and to bring the forgotten into language. The dragon, then, is not only a symbol of fantasy but a vestige of real, ancient encounters with a world that once was—and that still, in fragmented form, lies beneath our feet.

Fire as Divine Punishment

In the mythic imagination, fire is rarely neutral. It is either the force that illuminates and creates, or the power that consumes, punishes, and ends. Nowhere is this polarity more vivid than in the figure of the dragon. Across diverse mythological systems, dragons often arrive with flame—raining fire from the sky, torching cities, devouring forests, and reducing civilisations to ash. These fiery dragons are not mere narrative embellishments. Their presence across time and culture points to a shared symbolic response to real events—volcanic eruptions, comet impacts, and catastrophic conflagrations—that traumatised communities and demanded mythic explanation. When fire descended from above or erupted from the Earth, it was interpreted as judgement. And the dragon, with its jaws of flame, became the personification of divine wrath.

To understand the dragon's fire not as fantasy but as encoded memory, we must begin with the relationship between early human societies and environmental catastrophe. Volcanic eruptions are among the most dramatic and violent events humans can witness. They combine noise, light, heat, and force on a scale that overwhelms ordinary comprehension. The 1883 eruption of Krakatoa generated sound waves that circled the planet multiple times. The pyroclastic flows of Vesuvius in 79 CE buried entire cities in seconds. But long before these historically documented events, prehistoric eruptions would have left indelible impressions on the people who witnessed them. The sudden appearance of fire and smoke from the ground, followed by darkness, famine, or disease, would not have been interpreted mechanistically. It would have been experienced as divine or demonic punishment. The fire-breathing dragon becomes, in this context, not a metaphor but a *model*—a symbolic code for overwhelming terrestrial fire.

One of the most persuasive examples of myth responding to volcanic trauma is found in Indonesia, where the eruption of Mount Tambora in 1815 produced what came to be known as the "Year

Without a Summer." The global climatic disruptions caused crop failures, mass starvation, and social unrest across continents. In the oral traditions of local peoples, the mountain's eruption was attributed to angry spirits, ancestral violations, and elemental imbalance. These interpretations did not arise from ignorance, but from necessity. Where scientific causation was inaccessible, symbolic causation filled the explanatory gap. The fire, to be integrated into cultural memory, needed a face. The dragon—or its regional equivalent—offered that face.

Ancient Mesoamerican mythologies also reflect this symbolic encoding. The Aztec god Xiuhcoatl, often rendered as a fire serpent, is both a celestial weapon and an agent of destruction. He appears in codices wielded by gods to bring drought, war, or punishment. His serpentine form links him to both sky and earth, suggesting a convergence of meteorological and seismic force. In this case, fire is not an arbitrary trait but a *cosmological verdict.* The burning serpent is the means by which divine justice descends upon a disordered world. This logic is mirrored in Hindu texts, where dragons like Vritra block rivers and bring drought, only to be slain by gods wielding thunder and flame. These myths do not merely tell stories; they record *interpretive frameworks* for natural disaster.

Fire, especially when emerging from the sky, also has astronomical associations. Throughout human history, comets, meteors, and fireballs have terrified onlookers. Unlike stars, these celestial events were unpredictable, fast-moving, and often accompanied by sonic booms or flaming trails. In cultures where the sky was regarded as a realm of order and balance, any rupture in its apparent harmony—especially a fiery one—was viewed as an omen. The Tunguska event in 1908, while recorded in a literate scientific age, produced folkloric reverberations among the Evenki people of Siberia. They interpreted the explosion, caused by a meteoric airburst, as the wrath of a sky deity. One can only imagine the impact of similar but undocumented events in prehistory. When fire fell from the heavens and scorched the Earth, it would not have taken long for stories of sky-serpents or flame-bearing beasts to become encoded in oral tradition.

The Younger Dryas Impact Hypothesis, though still debated, proposes that a large comet or fragmented bolide struck North America around 12,800 years ago, contributing to rapid climate shifts and widespread megafaunal extinction. Some researchers have suggested that myths describing fiery destruction, great floods, and the death of ancient animals may contain embedded memories of this event. In the traditions of the Ojibwe, for instance, there are accounts of "the long-tailed star" bringing fire and transformation. Similarly, the Zuni and Hopi speak of past worlds ended by fire from above. These motifs are not reducible to astronomy alone, but when triangulated with geological evidence, they suggest that fire myths may encode more than allegory—they may carry *ancestral data* filtered through symbolic logic.

Weaponised fire adds another layer. Long before gunpowder or industrial explosives, ancient warfare employed fire as both tactic and spectacle. Flaming arrows, oil-filled pots, and scorched-earth campaigns were designed to destroy not only infrastructure but morale. The fire-dragon becomes, in this light, a mythic amplification of technological fear. In Chinese and Indian epics, fire-bearing weapons are wielded by heroes and gods, often described in terms that blur the boundary between biology and machinery. In the *Mahābhārata*, for instance, divine missiles emit flames, heat, and blinding light—descriptions that eerily anticipate modern munitions. The *Agneyastra*, or "weapon of fire," is said to burn entire regions, mirroring in symbolic form the feared results of real conflagration.

Importantly, these motifs are not restricted to ancient texts. The modern imagination remains tethered to the same symbolic grammar. When nuclear weapons were first tested and deployed in the 20th century, many observers described the mushroom cloud as a dragon, a serpent, or a fire god. In Japanese accounts of Hiroshima, the blast was likened to a sun falling from the sky—divine punishment rendered technological. Science fiction and fantasy, far from breaking with tradition, continue to frame fire-breathing creatures as apex threats. From Godzilla's radioactive breath to Tolkien's Smaug, the fiery dragon continues to embody the terror of disproportionate, elemental force.

The cognitive durability of fire as divine punishment can be explained, at least in part, by the brain's relationship to fire itself. Neurological studies show that firelight produces a unique pattern of engagement—stimulating attention while lowering anxiety. Around fire, early humans gathered not only for warmth but for storytelling, ritual, and collective memory formation. But uncontrolled fire reverses that emotional relationship. Wildfires trigger deep fear responses. The sound, speed, and unpredictability of fire are neurologically primed to elicit panic. To assign agency to fire—especially malevolent agency—is to make it psychologically manageable. The dragon, in this regard, is not just a story. It is a symbolic firewall, a way to cognitively contain the otherwise overwhelming force of natural or human-made infernos.

There is also a moral dimension to fire myths. Fire rarely appears in myth as *mere* destruction. It is typically framed as a response—to hubris, disobedience, corruption, imbalance. The punishment is not random but *earned*. This moral framing allows societies to integrate trauma into a coherent worldview. When cities burn, forests fall, or communities vanish, the myth offers an explanation: the dragon came because we erred. This is not blame in the modern psychological sense; it is cosmological logic. It transforms catastrophe into narrative, and suffering into structure.

Modern societies, for all their scientific literacy, continue to operate within this symbolic economy. Climate change is increasingly framed in mythic terms—fires that rage out of season, skies turned red, species fleeing, deserts encroaching. The language of reckoning re-emerges. While the dragon may no longer have scales or wings, its symbolic function remains intact. It warns, it punishes, it brings fire where order has failed.

To examine the motif of dragon fire across civilisations, then, is not to catalogue fantasy but to trace a cognitive and cultural pattern. Fire, whether from volcanoes, the sky, or human hands, has always posed an existential threat. The mythic imagination responded not by denying the terror, but by shaping it into a form that could be remembered, retold, and re-integrated. The fire-breathing dragon is a cultural interface for experiences that exceed comprehension—a mythic synthesis of trauma, fear, awe, and moral instruction.

That it persists into the present is not evidence of myth's irrelevance, but of its enduring precision. Fire has always been more than a tool. It is a threshold: between survival and annihilation, between signal and catastrophe, between human agency and cosmic force. And when it crosses that threshold, we name it—not just to understand it, but to survive it. The dragon, wreathed in flame, continues to perform that task. Not as superstition, but as symbol. Not as escape, but as recognition. Fire speaks. Myth listens—and answers.

Neurobiology of Fear and Flight

The dragon may never have walked the Earth, but it has always lived in the mind. Its anatomy—a hybrid of wings, claws, scales, fangs, serpentine movement, and fire—is an evolutionary supercut of the features most likely to trigger human alarm systems. The widespread appearance of dragon-like creatures across unconnected cultures has long fascinated mythologists, historians, and folklorists. But the universality of this figure may be less about shared stories and more about shared neurology. The brain is not a blank slate; it is an evolved organ tuned to detect danger and respond to threat. And when the brain constructs symbols, especially enduring ones, it does so from the building blocks of survival. The dragon archetype is not a cultural accident—it is a cognitive artefact, rooted in the structure and function of the human fear system.

At the core of that system is the amygdala, an almond-shaped cluster of nuclei located deep within the temporal lobes. The amygdala acts as a kind of early warning device, constantly scanning sensory input for signs of threat. What makes it so efficient—and so vital to survival—is its speed. The amygdala can activate a fear response before the prefrontal cortex has even had time to process what is being perceived. This means that fear often precedes conscious recognition. You jump back from the shadow on the path *before* you realise it's only a stick. From an evolutionary perspective, this hypervigilance is adaptive. Mistaking a shadow for a snake is a low-cost error. Failing to identify a predator in time can be fatal.

This biological imperative has shaped not just behaviour, but perception itself. Human beings are evolutionarily biased toward over-detecting danger. We are more likely to perceive threat where none exists than to miss a real one. The visual system, especially the midbrain tectum, is primed to respond rapidly to movement patterns characteristic of predators—lurking, pouncing, slithering. Studies on visual cognition demonstrate that snake-like forms are among the fastest recognised shapes in the animal kingdom, and that this recognition occurs even in infants. Researchers like Lynne Isbell have argued that the primate visual system evolved specifically to detect snakes, whose presence represented a unique combination of stealth, suddenness, and lethality in ancestral environments.

But the fear circuitry is not restricted to the recognition of snakes. It is similarly attuned to other apex predators—big cats, raptors, and, notably, fire. Each of these elements contributes distinctively to the anatomy of the dragon. Serpentine movement evokes the primal unease associated with venomous snakes. Wings and talons mirror the threat posed by birds of prey. The capacity for breath that incinerates evokes the uncontrolled terror of wildfire, one of humanity's most ancient and universal dangers. The dragon is not just one threat—it is a convergence of many. Its form is an archetypal composite, constructed from the brain's most persistent danger models.

The flight-or-fight response, orchestrated by the autonomic nervous system, underpins our engagement with these perceived threats. When triggered, the hypothalamus initiates a cascade: adrenaline surges, the heart rate increases, pupils dilate, and energy is diverted from long-term functions like digestion to immediate muscular readiness. This physiological shift is so rapid that rational thought is often overridden. We are, in those moments, not analytical creatures but reactive ones. The dragon, as an imagined stimulus, functions as a perfect trigger—eliciting not just cognitive fear but full-body preparedness. Myths of dragon encounters often emphasise this totality of response: the trembling hero, the frozen villagers, the scorched land. These are not narrative embellishments. They reflect an accurate phenomenology of terror.

Fear, however, is not experienced in a vacuum. It is shaped by memory, context, and social learning. The hippocampus, which encodes memory, works closely with the amygdala to ensure that threats are remembered—and often exaggerated. The brain does not prioritise accurate recall; it prioritises effective avoidance. This helps explain why mythic threats tend to escalate over generations. A dangerous animal becomes a monster. A deadly fire becomes the wrath of a god. A serpent becomes a dragon. The symbol evolves in tandem with the brain's need to generalise and remember. This memory inflation is not distortion—it is a safety feature.

Interestingly, the same neural systems that govern fear are also involved in curiosity and learning. The mesolimbic dopamine pathway, which mediates reward, overlaps with circuits that respond to threat. This overlap creates a paradox: we are drawn to that which scares us. This explains why children are fascinated by dinosaurs, why horror films attract millions, and why dragon myths persist even when belief in them has faded. The dragon engages multiple systems simultaneously—arousing fear, stimulating attention, and encoding memory. It is neurologically sticky.

This stickiness is reinforced by cultural transmission. From a memetic perspective, the dragon is a high-fidelity symbol—easy to reproduce, emotionally charged, and richly adaptable. It can be terrifying or noble, chaotic or orderly, destructive or protective. Its flexibility enhances its transmissibility. But the emotional core remains constant: dragons evoke awe. And awe, neuroscientifically speaking, is a compound state that combines fear and admiration. It activates the default mode network (associated with self-referential thought) and the salience network (associated with attentional shifts), creating a temporary reconfiguration of the brain's processing hierarchy. In this state, humans are more susceptible to influence, more open to learning, and more likely to remember what they experience. Myths that induce awe are, therefore, not just compelling—they are *neurologically privileged.*

There is also an evolutionary rationale for these mythic codings. The young of social species must learn rapidly what to avoid, and narratives are an efficient tool for this purpose. Telling a child, "Don't go near the river, there are snakes," is useful. But telling a child, "In the river lives the dragon who devours the disobedient," is more memorable. Myths are not lies. They are pedagogical technologies—compact, emotionally loaded vehicles for transmitting adaptive behaviour. The dragon, in this view, is a didactic threat: a mnemonic device designed to keep people alive. Its fire teaches caution. Its wings teach respect for power. Its presence near caves or mountains marks danger zones to be navigated carefully.

The neurobiology of fear also helps explain why dragon myths resist obsolescence. Scientific advances may have replaced the literal belief in monsters, but they have not dismantled the circuitry that made those beliefs effective. The structures that once responded to dragons now respond to symbolic equivalents—global threats, technological dangers, ecological crises. We still imagine threats as looming, hybrid, and devastating. The dragon has morphed into the metaphorical: financial crashes, pandemics, artificial intelligence run amok. But the architecture of alarm remains unchanged. We give these threats form because our brains are not designed to fear abstractions. They are designed to fear patterns, bodies, and trajectories. The dragon continues to offer all three.

Cognitive neuroscientist Joseph LeDoux, in his work on the emotional brain, emphasises the importance of distinguishing between the *experience* of fear and the *mechanism* of fear. One is conscious, the other unconscious. Myths like the dragon integrate both. They externalise the mechanism in a form that can be consciously engaged. By giving the unconscious terror a shape, they allow cultures to narrativise and ritualise it. The hero's confrontation with the dragon, then, is not just symbolic. It mirrors the neurological act of bringing amygdala-generated fear into conscious regulation—a movement from limbic reaction to cortical control. Myth, in this context, becomes a tool for managing neurobiology.

This management is not only individual but social. Collective rituals—festivals, dances, storytelling, or reenactments—transform fear into shared meaning. In many traditions, dragon myths are embedded in annual ceremonies that ritually re-enact the defeat of chaos. These are not merely celebrations. They are symbolic recalibrations of group cohesion in the face of remembered danger. The neurobiology of fear does not exist in isolation. It is co-regulated through culture. And myth provides the scaffolding for that regulation.

To read the dragon through the lens of neurobiology is not to diminish its power but to clarify it. The dragon is not simply a fantastical chimera; it is a crystallised expression of how our nervous systems have evolved to encode, respond to, and socially transmit threat. Its wings may not flap in the real world, but its shape is drawn from the architecture of our most ancient fears. In that sense, the dragon is not a mythological animal. It is a cognitive pattern—burned into the nervous system long before it was inked onto parchment. It endures not because it is unreal, but because it is *biologically plausible.* And it teaches us, even now, how deeply the line between story and survival is written in the body.

Chapter 3
The Great Flood and the Geological Imagination

Water remembers nothing, yet somehow, it remembers everything. It moves without sentiment, carrying, swallowing, scouring; it destroys with the same ease that it preserves. In one gesture, it wipes away the surface traces of time; in another, it locks fragments of the past into its quiet depths. Across cultures and centuries, the flood emerges as an enduring paradox—both apocalypse and genesis, feared as retribution, embraced as renewal. And what is most curious is not merely that so many ancient peoples told stories of vast, world-altering deluges, but that they did so almost everywhere, with striking similarities in imagery, structure, and moral framing. These are not scattered inventions. They are recurring shapes in the human imagination, patterns that cross oceans and languages. From the riverbanks of Sumer to the islands of the South Pacific, from Mesoamerican plateaus to Siberian tundra, the flood motif rises and falls through human myth like a memory that refuses to remain buried. But might these memories rest on something more solid than metaphor—on the very bedrock of the Earth's history?

Our modern instincts tell us to file myth under allegory, at best, and invention, at worst. That, perhaps, sells our ancestors short. Oral cultures—particularly those without written records—did not tell stories solely for diversion or philosophical reflection. Myths were also mnemonic devices, cultural storage systems for knowledge that mattered. A flood story, told and retold for centuries, could serve not only as a moral tale, but as a practical reminder: water can rise, and it will again. In this light, certain flood narratives begin to look less like whimsical storytelling and more like coded geological reports, passed forward in the only medium available—symbol and ritualised plot.

Consider the Mesopotamian *Epic of Gilgamesh*, one of the oldest literary artefacts we possess. It recounts how Utnapishtim, forewarned by divine messengers, constructed a massive vessel to survive a deluge that swept away the known world. His tale precedes the Hebrew account of Noah by centuries, yet the bones of the two stories are nearly identical: a divine warning, the preservation of life in miniature, the overwhelming flood, the dawning of a cleansed world. The same elements appear in the oral histories of the Hopi, the Yoruba, the Tamil, and the Aboriginal peoples of Australia. Geography, theology, and language shift, yet the skeleton of the tale remains recognisable. Such persistence begs for explanation.

One proposal with significant scholarly weight is the Black Sea deluge hypothesis. In 1997, marine geologists William Ryan and Walter Pitman suggested that around 5600 BCE, rising Mediterranean waters breached the Bosporus, sending torrents into the Black Sea basin and raising its level dramatically in a matter of months or years. To those living along its then-fertile shores, it would have been as if the sea itself had turned predator—sweeping in without warning, claiming fields, homes, and lives. If the memory of such a cataclysm endured in oral form, it could easily have seeded the flood narratives of the Near East, later crystallising into the Mesopotamian and biblical accounts. This is not just a historical curiosity; it is a reminder that myth may preserve the emotional sediment of ancient disaster.

Yet the Black Sea episode might be only one layer in a much deeper stratigraphy of human memory. At the close of the last Ice Age, roughly 11,700 years ago, the Earth lurched abruptly out of the Younger Dryas, a period of intense cold whose origin remains debated. During the ensuing transition, a dramatic episode known as Meltwater Pulse 1B saw sea levels climb by as much as twenty metres within five centuries—possibly faster in some regions. For the coastal peoples of that age, such an onrush would have been nothing short of annihilation. No maps, no forecasts—only the horizon swallowing the land. Events like this would leave their imprint, not in the written record, but in the minds of survivors and the stories they told their children.

Seen this way, flood myths are not the residue of a single disaster, but the distillation of many—compressed, over generations, into archetypal form. Like the layered deposits in an ice core, these myths may contain traces of multiple inundations, each reframed to fit the cultural lens of its tellers. The consistency across continents is not proof of invention, but of convergence: faced with the same elemental threat, human societies responded with remarkably similar symbolic architectures.

Why, though, does water occupy this singular place in our imagination? Fire devastates, but water purifies. Earth trembles, but water envelops. Its violence is patient, smothering rather than striking. To the psyche, it communicates helplessness; to the collective mind, it signals an ending. The philosopher Gaston Bachelard called water "the substance of loss," a medium that erases even as it records. In the flood story, this becomes the engine of renewal: worlds must vanish for others to appear. In its most distilled form, the flood is the cosmic reset button.

Nearly every variant frames the deluge as judgement. The world before is corrupt, prideful, disordered; the flood sweeps the slate clean. Whether the offence is hubris in Greek myth, taboo-breaking in Polynesian tradition, or moral decay in the Near East, the pattern holds. The moral weight is crucial—without it, the flood is a meaningless accident, likely to fade from memory. With it, the event becomes a lesson, ensuring its survival through generations.

The mechanism is still visible in modern times. When Hurricane Katrina struck New Orleans in 2005, reportage quickly took on mythic overtones. For some, it was divine chastisement; for others, an indictment of political negligence. The 2004 Indian Ocean tsunami and the creeping encroachments of climate-driven sea rise have evoked similar reframing. The physical waters recede, but the symbolic waters linger, reshaping public consciousness long after the event itself.

To ritualise trauma is to safeguard against forgetting. This instinct—turning ecological catastrophe into narrative—is not primitive; it is adaptive. Stories can carry warnings long after dates and measurements have vanished. And in our present century, with seas rising again and coastal cities under threat, those ancient warnings seem less like relics and more like prophecies. Science tells us how fast the water will come; myth tells us what it will mean when it does.

Flood myths, then, cannot be neatly categorised as either history or allegory. They are both, and more besides: geological testimony embedded in cultural DNA; emotional archives in the guise of parable; philosophical reflections on endings and beginnings. And, like the waters they describe, they are not still. As once before, the tides are shifting. The question is whether we will read the signs in time—not only in the satellite data, but in the stories we have been telling for thousands of years.

The Flood in Every Culture From the Epic

It is rare to find a motif that unites humanity across language, landscape, and epoch. Yet the flood myth does just that. From the riverbanks of Mesopotamia to the rainforests of Brazil, from Arctic tundra to the Pacific archipelagos, one narrative rises like a tidal swell through human history: the world was drowned, and only a few survived. This uncanny recurrence is not simply an anthropological curiosity—it is a signal, a pattern etched into the cultural DNA of our species. The sheer consistency of flood myths across disparate civilisations suggests something deeper than coincidence, more enduring than fantasy. They point to a shared ancestral trauma—one encoded in story, ritual, and memory.

The Sumerian *Epic of Gilgamesh*, composed around 2100 BCE, offers the oldest written version of a deluge narrative. In it, the god Enlil grows weary of humanity's noise and sends a flood to erase them. The sage Utnapishtim is warned by the god Ea and builds a massive boat to preserve life. This story predates, yet closely parallels, the more widely known narrative of Noah in the Hebrew Bible's *Book of Genesis*. In both, a divine decision leads to catastrophe; in both, a single righteous man preserves biodiversity through an ark; and in both, the waters recede to offer humanity a second beginning. The parallels are not just structural—they are specific, down to the sequence of events, the use of birds to test dry land, and the final resting of the boat on a mountain.

What's striking is that this structural consistency appears far beyond the ancient Near East. In the Andean highlands, the Incas told of a time when the god Viracocha caused a flood to destroy a race of giants, sparing only a chosen pair. In Hindu cosmology, *Manu*, the first man, is forewarned by the fish-avatar of Vishnu about a deluge that will cleanse the world; he builds a boat and survives. In Australian Aboriginal traditions, stories speak of floods that reshaped the land and wiped out earlier beings, sometimes seen as ancestral spirits or creators. These stories emerge from cultures that had no known contact with one another, no shared religious canon, and no reason to independently develop nearly identical mythic sequences—unless they were all responding to something real.

This invites an uncomfortable but intellectually necessary question: could the flood myths encode actual environmental memory? Oral tradition, far from being fiction, functions in many indigenous societies as a tool of long-range historical continuity. Studies of Aboriginal songlines, for instance, have shown that oral maps transmitted for millennia accurately describe features of landscapes submerged after the last glacial maximum. When interpreted through this lens, flood myths become not allegory but memory—refracted through symbolic narrative, but rooted in observation and experience.

Consider the accounts of the Native American Ojibwe, who speak of a time when the Earth was flooded and the surviving animals recreated land atop a turtle's back. Though the tale is saturated with cosmological symbolism, its underlying premise—a sudden, devastating deluge that reorders the known world—mirrors narratives found in Asia, Africa, and Oceania. The Yoruba people of West Africa recount that the god Olokun, deity of the sea, once flooded the land in anger, forcing other deities to restore balance. Among the Chinese Miao, a great flood narrative involves two siblings, Fu Xi and Nuwa, who survive and repopulate the Earth, echoing a pattern seen across Eurasian traditions.

The tendency to reduce these myths to "primitive explanations for weather" fundamentally underestimates the depth of human cognition and symbolic reasoning. These are not haphazard tales told to explain a thunderstorm; they are mythic frameworks for processing societal collapse, radical ecological transformation, and civilisational reset. They carry moral gravity and cosmological insight. In many cases, the flood follows a pattern of human hubris or imbalance—noise, wickedness, overpopulation—prompting a divine or cosmic reset. The deluge is not just destruction; it is judgement. And more importantly, it is renewal.

That said, not all scholars agree on the historical referent. Some argue that the prevalence of flood myths owes more to universal psychology than to shared memory of a real event. The psychoanalyst Carl Jung viewed the flood as an archetype of the unconscious—symbolising the inundation of the ego by repressed forces. Mircea Eliade interpreted floods as symbols of the return to a primordial chaos, a cleansing of the temporal world in order to restore sacred order. From this angle, the myth is less geological record and more spiritual metaphor. The universality of the flood motif, then, stems not from shared history but from shared mind.

But these interpretations are not mutually exclusive. It is plausible—indeed, likely—that a real sequence of cataclysmic events became psychologically charged, morphing through centuries into myth. Catastrophic sea level rise during Meltwater Pulse 1A and 1B—events associated with the end of the last Ice Age—transformed coastlines globally. Whole societies would have been displaced as once-habitable land disappeared beneath rising seas. For those communities, the ocean would not have been metaphor. It would have been devouring. Their stories would naturally carry the emotional residue of survival, fear, and loss. Over time, these residues crystallised into mythic form.

Furthermore, the consistency of motifs—ark-building, animal preservation, mountain landings—points to more than independent invention. It suggests that certain key survival strategies were widely known and mythologised. Building a floating vessel, preserving seed and livestock, seeking high ground—these are not just narrative flourishes but practical adaptations. If retold and ritualised across generations, they become encoded in culture not as history but as sacred instruction. Myth, in this context, is not delusion. It is adaptive memory.

The flood also functions as a civilisational mirror. It reflects how a culture interprets crisis, assigns blame, and envisions redemption. In some traditions, the flood is punishment for human moral failure; in others, it is a cosmic cycle beyond human control. The Abrahamic flood punishes sin; the Hindu flood resets the *yuga* cycle; the Polynesian flood reflects environmental imbalance. These interpretative differences matter. They reveal how societies assign agency to catastrophe—whether to deities, natural forces, or human hubris—and thereby shape responses to future risk.

Modern flood narratives, while shorn of divine agents, still echo these ancient forms. Climate change-induced sea level rise is framed as the result of our own excess—carbon emissions, overconsumption, disregard for ecological balance. The narrative structure remains: wrongdoing leads to inundation, followed (we hope) by correction. Even our films—*The Day After Tomorrow*, *Noah*, *2012*—reprise this mythic arc. The settings have changed, the agents are now climatologists or computer models, but the essential story persists. The flood as symbol has simply migrated into scientific discourse and cinematic spectacle.

What this suggests is that the flood myth is not a static relic of ancient cosmology but a living framework through which humans continue to process ecological disruption. It remains potent precisely because it is layered: it is simultaneously historical, moral, emotional, and ecological. Dismissing it as fantasy is not only intellectually lazy—it is culturally myopic. These stories deserve re-reading, not as quaint folktales, but as encrypted data sets—carrying warnings, instructions, and reflections from deep time.

It is no accident that in so many of these myths, survival hinges on listening. A god whispers, a vision warns, a dream instructs. The one who hears and acts is the one who endures. This metaphor remains resonant. Whether we are listening to satellite data, indigenous knowledge, or ancient myth, the challenge is the same: to hear the rising water not only as physical threat, but as meaning. The flood may return—not as punishment, but as proof that the old stories still speak. And the question, then as now, is whether we are ready to listen.

Black Sea Deluge Theory Scientific evidence

When scholars William Ryan and Walter Pitman published their groundbreaking hypothesis in the late 1990s, suggesting that the Black Sea underwent a catastrophic flooding event around 5600 BCE, the idea sent a ripple through both the geological and mythological communities. Not merely because it offered a compelling case of sudden environmental upheaval, but because it reopened a long-standing question: do our oldest flood stories contain an echo of real geological memory? Their research—interdisciplinary, data-driven, and evocative—proposed that the sudden influx of Mediterranean waters into the Black Sea basin might have been the historical catalyst for flood myths such as that of Noah, Utnapishtim, and others that seem to converge across time and civilisations. The Black Sea Deluge Theory stands at the intersection of archaeology, paleoclimatology, and myth, posing a provocative idea—that geology and oral tradition are not strangers but estranged siblings.

The basic outline of the theory is deceptively simple: at the end of the last Ice Age, global sea levels were still rising due to the melting of vast glacial sheets. The Mediterranean had already filled to modern levels, but the Black Sea was a largely isolated freshwater lake, its surface far lower than that of the Mediterranean. According to Ryan and Pitman's model, the narrow Bosporus Strait acted as a natural dam—until, at a certain threshold, it was breached. What followed, they argue, was not a slow seepage but a violent, cascading rush of saltwater from the Mediterranean into the Black Sea basin. The result would have been cataclysmic for Neolithic communities living along its shores. Some estimates suggest that up to 60,000 square miles of habitable land could have been inundated within a few years—or even months—forcing massive displacements of human populations and reshaping the ecological and cultural landscape of the region.

What makes this scenario particularly compelling is the archaeological context. Excavations around the Black Sea, particularly in Bulgaria, Ukraine, and Turkey, have revealed submerged structures, traces of human habitation below current sea levels, and abrupt shifts in sedimentation patterns consistent with a large-scale marine incursion. While the exact timing and magnitude of the flooding remain debated, few now doubt that a significant transformation of the Black Sea region occurred during the Holocene, and that human witnesses may well have transmitted the event through story.

This potential connection between geological catastrophe and mythic narrative becomes especially intriguing when viewed through the lens of cultural diffusion. The Near East was a crucible of early civilisations, and the Fertile Crescent—home to the Sumerians, Akkadians, and later Hebrews—lies within proximity to the Black Sea's influence zone. If Neolithic peoples fled the rising waters, migrating southward with their memories intact, it is plausible that those memories took form in the epic cosmologies that later emerged in Mesopotamian and Levantine traditions. This is not to suggest a simplistic one-to-one correlation—Noah did not sail the Black Sea, and Utnapishtim did not reside in modern Bulgaria—but rather that the structural imprint of an unforgettable event shaped the psychological and narrative frameworks of multiple cultures over time.

Critics have challenged aspects of the theory, particularly its reliance on extrapolated sea-level data and assumptions about the rate of the flooding. Subsequent studies have produced alternative timelines, suggesting a more gradual marine transgression rather than a catastrophic breach. Some geologists argue that the stratigraphy of the Bosporus Strait shows evidence of a complex, multi-stage flooding process, rather than a single, decisive event. Yet even if the flood was slower than originally proposed, the implications for human experience may remain profound. What matters for cultural memory is not just the physical rate of change, but the perceptual rupture it induces. If within a generation or two, communities found their arable land submerged, their settlements abandoned, and their sacred sites lost to the sea, the psychological impact would be comparable to catastrophe. A story of rising water becomes, inevitably, a story of divine retribution, of cosmic reset, of the boundary between order and chaos.

This transition from event to myth is not passive—it is interpretive, moralising, and mnemonic. Myths do not merely remember; they teach. The narrative arc of the flood—as punishment, as trial, as survival—becomes a cognitive template. In this context, the Black Sea Deluge Theory is not simply a palaeogeographic argument; it is an insight into how civilisations metabolise trauma. The event becomes storied, storied becomes sacred, and sacred becomes foundational. For cultures rebuilding on the far side of disaster, the mythic flood is a means of re-establishing cosmic order. It explains the unexplainable, justifies the loss, and affirms the survivor.

Furthermore, this model offers a critical counterpoint to simplistic dichotomies between myth and science. The assumption that ancient people "invented" flood myths to explain weather or crop failure fails to account for the accuracy and endurance of these motifs. If the Black Sea event seeded oral traditions that evolved into the Mesopotamian, Biblical, and even Greco-Roman deluge accounts, then myth is not opposed to evidence—it is evidence, preserved in symbolic code. Much like the way Polynesian navigators used stars and oral chants to map thousands of miles of open ocean, ancient flood myths may have encoded real historical knowledge in metaphorical form.

Yet the theory also opens deeper ethical and epistemological questions. If we accept that a natural disaster could ripple through culture for millennia, altering not only demographic patterns but cosmological worldviews, what does that imply about our present-day vulnerabilities? As we face rising seas, melting glaciers, and volatile climate systems, are we in the early stages of a future mythology in the making? Will the survivors of future catastrophes craft their own deluge epics—ones that begin not with gods but with satellites, not with oracles but with ignored warnings? The Black Sea

Deluge Theory forces us to reflect on the longue durée of human memory—how we frame, forget, or mythologise what we cannot fully understand.

It also reminds us of the humility required in interpreting the past. Mythology is not failed history; it is parallel knowledge. To dismiss it is to cut ourselves off from the psychological and cultural software that helped previous generations endure collapse. Whether or not the Black Sea flood was the single origin of global flood myths—a claim that most scholars treat with scepticism—it stands as a powerful candidate for how one regional trauma could have influenced a wide array of mythic traditions. It invites us to think not in binaries—true or false, history or fiction—but in gradations of symbolic resonance.

And therein lies the real value of the theory. It doesn't need to be definitive to be illuminating. It shows that science and story are not strangers but neighbours. The sediments of the Black Sea and the lines of the *Epic of Gilgamesh* may, in their own ways, be describing the same thing: a boundary crossed, a world drowned, a memory that refuses to die. In that confluence of flood and meaning, of basin and book, we glimpse a model for rethinking what it means to remember, and what it means to survive.

Echoes of Meltwater Pulse 1B Global sea

Roughly 11,700 years ago, the Earth emerged from the tail end of a deep freeze—a climactic period known as the Younger Dryas. It marked not only the transition from the Pleistocene to the Holocene, but also one of the most dramatic hydrological episodes in human prehistory: Meltwater Pulse 1B. During this interval, sea levels rose precipitously, with estimates suggesting increases of up to 28 metres within just a few centuries. This wasn't a slow, generational creep; it was a profound rupture in the stability of coastlines, ecosystems, and the rhythm of human settlement. For any coastal community alive at the time—and archaeological evidence suggests many were—this change would have felt biblical. Except the Bible, of course, wouldn't be written for thousands of years. What these people did have was oral tradition. And what those traditions have preserved, scattered across the globe in eerily similar forms, are stories of floods, deluges, submerged lands, and cataclysmic waters sent by gods, spirits, or angry skies. The question is not whether these stories reflect reality. The question is: how much?

Meltwater Pulse 1B is not speculative. It is robustly supported by paleo-sea level reconstructions, ice core data, coral reef analysis, and glacio-isostatic modelling. Its triggers remain debated, but its existence is not. What's intriguing is the extent to which this real and measurable event coincides with the hypothesised Younger Dryas impact—a proposed comet or asteroid collision that, according to proponents, may have precipitated temporary global cooling followed by sudden warming and ice sheet destabilisation. Though the impact hypothesis remains controversial, the idea that a celestial event could have cascaded into hydrological upheaval is consistent with both mythic narratives and climatological models. In this convergence of sky and sea, geology and story, we encounter the kind of structural resonance that myths seem built to carry.

Consider the mythic data. Indigenous Australian accounts speak of the ocean rising to swallow the land, an event their Dreamtime stories mark not only as a historical occurrence but a moral one. In India, the Matsya Purana describes a divine fish warning the sage Manu of an impending flood, urging him to preserve the sacred texts and seeds of life. The Maya Popol Vuh recounts cycles of creation and destruction, including a great flood sent to obliterate earlier, failed human prototypes. And in Dogon cosmology from West Africa, the origin stories weave together water, stars, and catastrophe with remarkable symbolic sophistication. What unites these traditions is not only the theme of water's wrath, but the structuring of that water as both erasure and renewal—a principle that aligns seamlessly with the geological record of Meltwater Pulse 1B.

One of the more persuasive theories linking oral tradition to palaeoclimatic events comes from linguist and mythographer Patrick Nunn, who has spent decades examining Pacific islander stories that speak of land loss, submerged villages, and disappearing coastlines. His research suggests that some of these stories preserve accurate geomorphological information that could date back over 7,000 years. If islanders could encode such changes into story over millennia, is it not plausible that the much earlier, much larger meltwater pulses—

particularly 1B—would find equally tenacious expression in the myths of cultures bordering the Atlantic, Indian Ocean, or what would later become the Mediterranean basin?

Importantly, storytelling is not simply about remembering. It is about encoding experience in a way that allows it to survive transmission, fragmentation, and reinterpretation. Myths are designed to be retold, remixed, and ritualised. In doing so, they can retain core thematic elements while adapting their details to the context of the culture passing them on. If the essential truth is that "the waters came, and everything changed," then that seed can grow into a thousand local variations—Noah, Deucalion, Atrahasis, Manu, Nu'u—each reflecting the narrative priorities of its audience while preserving a deeper ecological memory. What the Meltwater Pulse model allows us to do is reconnect that deeper memory to physical processes we can now measure and map.

Yet the implications go further than aligning myth with event. They challenge our understanding of time and memory itself. The standard assumption in much of modern historiography is that accurate memory decays rapidly. Oral cultures, however, often build systems for long-term information storage that are functionally mnemonic rather than textual. Songlines, initiation rituals, architectural alignments, seasonal festivals—all these act as memory palaces that encode astronomical, ecological, and even geological data. When glacial cycles affect not just the temperature of a region but its very topography—when seas rise to erase coastlines, force migration, or reshape river systems—the stories that emerge are not decorative. They are maps. And as any navigator knows, a map's utility lies not in its realism but in its fidelity to key patterns.

Meltwater Pulse 1B also forces us to rethink the speed of change. We often view climate shifts as either glacially slow or apocalyptically fast. But this event resists that binary. It was abrupt enough to be traumatic, yet prolonged enough to allow for adaptation. This in-between quality may explain why the myths that emerged are neither simple tales of destruction nor complete fables of divine rage—they are often ambivalent, cyclical, and complex. The flood is both punishment and opportunity, both cleansing and loss. That ambiguity matches the lived experience of communities who would have seen not a single cataclysmic wave, but a steady, terrifying erosion of the familiar: fields turning to marshes, freshwater wells growing brackish, sacred hills becoming islands.

Furthermore, the timing of Meltwater Pulse 1B—around the onset of the Holocene—places it at the dawn of agriculture, permanent settlement, and early state formation. It is precisely this period in which we begin to see the emergence of what might be called "structured myth": stories not just of spirits and animals, but of gods, kings, contracts, and cosmological ordering. If the trauma of sea-level rise provided not only a memory to preserve but a pressure to organise society anew, then perhaps myth became the software update for cultures undergoing unprecedented change. It offered a shared framework to explain, ritualise, and morally evaluate events that no individual could comprehend alone.

Debates continue over whether the Younger Dryas impact actually occurred, or whether Meltwater Pulse 1B was globally synchronous or regionally variable. But even if the specifics remain unsettled, the broader principle holds: the Earth's climate system has generated episodes of profound instability, and human beings have always responded not merely with tools, but with stories. Those stories are not distractions from history—they are its scaffolding.

Today, as we face our own phase of accelerating sea-level rise, this scaffolding becomes not just a record but a warning. What we dismiss as ancient superstition may, in fact, be the long tail of planetary feedback, whispered from the past in allegorical tones. The lesson is not that myths are factually accurate; it's that they are structurally informative. They teach us what to pay attention to, what to fear, what to cherish, and what to rebuild.

Meltwater Pulse 1B is not merely a data point in the paleoclimatic record. It is a candidate for a global memory scar—one etched not just in coral reefs or sediment cores, but in the shape of our stories, the structure of our myths, and perhaps even the architecture of our rituals. By examining it through the lens of both science and narrative, we do not reduce myth to geology or inflate geology into prophecy. We respect the possibility that deep human time is accessible not only through fossils and satellites, but through symbols—if we know how to read them.

Water as Symbol of Memory and Erasure Philosophically

Water has always signified more than just sustenance or destruction. Across cultures, disciplines, and centuries, it has consistently emerged as one of the most potent metaphors in the human symbolic repertoire—at once a force of annihilation and purification, of death and rebirth. What makes water so singularly resonant is not merely its physical omnipresence but its cognitive elasticity. It carries memory while also dissolving it. It nourishes civilisations and sweeps them away. In myth, ritual, and philosophical reflection, water becomes a liminal agent—neither entirely here nor gone, neither fully graspable nor inert. It is the element that marks thresholds: between chaos and order, between the past and the possibility of beginning again.

This ambivalence is central to the way water is used to process trauma in cultural consciousness. From the Mesopotamian deluge in the *Epic of Gilgamesh* to the Christian baptismal font, from the great flood of Manu in Hindu cosmology to the purifying ablutions of Islamic wudu, water is never just a medium. It is a message: of cleansing, of surrender, of starting over. More than that, it serves as a container for collective memory while also being the agent through which that memory is periodically washed away. It symbolises continuity and rupture, storage and deletion. To understand why this duality matters, one must appreciate that human cultures have long needed mechanisms for both remembering and forgetting—particularly in the aftermath of civilisational rupture.

At the neurological level, trauma imprints itself into memory with particular intensity. The amygdala, involved in emotional processing, works in conjunction with the hippocampus to encode vivid, sometimes intrusive recollections. But cultural trauma—on the scale of flood myths, invasions, or famines—demands a different kind of storage. It cannot reside solely in the individual brain. Instead, it becomes embedded in ritual, architecture, oral narrative, and landscape. Water, with its capacity to physically transform the environment while symbolically framing it, becomes a key agent in this encoding. When floods swallow settlements, when rivers change course, when islands vanish beneath rising seas, the event is remembered through narrative—but often mythologised through water's role as both destroyer and cleanser. This is not incidental; it is structurally necessary. It enables a community to process loss without being paralysed by it.

The philosophical potency of water as a metaphor for erasure lies in its formlessness. Unlike fire, which obliterates by reduction, or wind, which scatters, water absorbs. It takes in the past without leaving a trace, like ink diluted into a vast basin. This makes it the perfect analogue for historical trauma that is remembered not in the detail, but in the residue. A submerged city may leave behind artefacts and foundations, but in the cultural imagination, what often remains is the sense of something precious lost beneath the surface—something unreachable, unnameable, yet emotionally potent. Atlantis, Kumari Kandam, Ys, and Cantre'r Gwaelod are just some of the submerged civilisations remembered not as museum exhibits but as cultural aches. Their waters don't just erase land—they erase certainty, history, and continuity. And yet, they also give rise to longing, to myth, and to the imaginative reconstruction of lost golden ages.

What is striking is how frequently water is used not just to mark catastrophe but to authorise renewal. In Jewish tradition, the mikveh functions as a ritual bath of spiritual reset. In many African and Afro-Caribbean religions, water spirits preside over fertility, healing, and initiation. Christian baptism, often misunderstood as merely symbolic, functions at a deeper level as a theological rebirth—the old self 'dying' in water and the new emerging. Across these examples, water is not simply a backdrop. It performs a psychosocial operation. It marks a threshold from which new identity can emerge—individual or collective. And crucially, it enables forgetting—not in the sense of denial, but in the sense of letting go. As philosopher Paul Ricœur argued, forgetting is not the opposite of memory, but its necessary counterpart. No narrative can sustain itself without omission, editing, and silence. Water offers a way of doing this that feels natural rather than imposed.

This brings us to the structural role of water in mythology. Deluge myths are rarely just about punishment. They are about recalibration. The pre-flood world is almost always corrupt, excessive, chaotic. The flood comes not just as divine anger, but as a purging—allowing a new order to be born. What follows is not just survival but reconstruction. Noah plants a vineyard. Deucalion and Pyrrha repopulate the world. Manu discovers dharma in the post-deluge world. The waters do not simply end history—they reboot it. And this, in turn, reinforces water's position as a cultural tool for trauma processing. It enables what psychoanalyst Dori Laub might call "witnessing"—not just of what happened, but of the possibility of meaning after what happened.

Moreover, water's symbolic flexibility allows it to be both passive and active, external and internal. In literature and philosophy, this is exploited with deliberate precision. Heraclitus' famous assertion that "you cannot step into the same river twice" encapsulates not just the fluidity of experience, but the impossibility of perfect recollection. The river flows onward; memory changes with it. In Buddhist thought, too, water serves as a metaphor for the mind: flowing, reflective, capable of both clarity and disturbance. These frameworks share a common intuition: that memory is not fixed, and neither is history. What water offers is a way to hold that instability without denying it.

At a more grounded level, the ritual use of water often maps directly onto ecological memory. In communities living near floodplains, the seasonal rhythms of inundation and retreat become ritualised not just as agricultural patterns but as metaphysical events. The annual flooding of the Nile was not merely a hydrological cycle—it was the breath of the god Hapi, the sustainer of life. To forget the flood would be to forget the gift; to ignore its excesses would be to risk destruction. Here, water is not metaphorical—it is mythic in the most literal sense: a natural force made meaningful through narrative. It carries memory in the soil, the silt, the survival strategy.

In the modern age, our relationship with water has shifted. Dams, desalination, irrigation, and climate change have made us engineers of its flow. But this has not displaced its symbolic power. When cities flood today—Venice, New York, Jakarta—we don't just see infrastructure failure. We glimpse the return of something older: the idea that water always comes back. That nothing is ever permanently held at bay. In the face of rising seas, melting glaciers, and intensifying storms, water once again becomes a metaphor for reckoning. And just as ancient cultures turned to myth to process their flood events, we turn to science fiction, speculative fiction, and climate narratives to frame ours. The medium has changed, but the function remains.

Finally, the enduring power of water as both memory and erasure lies in its ethical charge. It demands of us a decision: what to carry forward, and what to release. Trauma, after all, is not just what happens—it is what is remembered without integration. Water, in its symbolic and ritual forms, offers the possibility of that integration. It allows cultures to grieve without being consumed, to recall without being paralysed. The flood myth, far from being an escapist fantasy, is a sophisticated cultural mechanism for surviving what otherwise might have destroyed us. It gives form to chaos, meaning to loss, and permission to begin again.

In reading water this way—as both the grave and the cradle of collective memory—we glimpse not only the symbolic intelligence of our ancestors, but the resilience of myth as a technology of continuity. The flood that drowns the old world does not silence the story. It makes space for a new one to be told.

Chapter 4
The Sky Knows Everything

Before temples rose from the earth, there were only stars. Before alphabets etched human thought into permanence, constellations had already sketched their own narratives across the night. Across epochs and civilisations, the sky was never a mute expanse—it was a living ledger, inscribed with meaning, order, and guidance. For ancient peoples, looking upward was never idleness; it was a discipline, a necessity. You read the heavens as you might a calendar or a clock—calculating when to plant or harvest, when to mourn or rejoice. Long before the codification of astronomy, myth served as the scaffolding for a kind of proto-science, rendering the cosmos intelligible through ritual, symbol, and story. In those tales, the heavens did not merely shine; they spoke. And their language, remarkably, has outlasted the very cultures that first gave it voice.

Here we follow a provocative yet increasingly credible proposition: that myth and sky were never separate spheres, and that many of humanity's oldest tales may not be poetic fictions at all, but carefully crafted memory-tools for astronomical truths. Think of Orion—Osiris to the Egyptians—or the Pleiades, who appear as the Seven Sisters in Greek, Aboriginal, and Native American traditions. The recurrence of such patterns hints at an ancient literacy—one not bound to parchment, but to celestial cycles, angular relationships, and the dependable rhythm of recurrence. Constellations were calendars. Their seasonal shifts were signals. Their legends? Instruction manuals for living in sync with a cosmos on which survival depended.

The fact that peoples separated by oceans and millennia fastened their attention on the same clusters of stars is no statistical accident to be brushed aside; it is a pattern worth unpicking. Few examples are more compelling than the Pleiades—visible in both hemispheres, they appear in Japan as Subaru, to the Māori as Matariki, in Greece as the Seven Sisters, and in Cherokee lore as sisters pursued or hidden. Always, the stars are tied to agricultural timing. Such uncanny convergence invites questions. Did different societies, by virtue of shared cognitive wiring, independently craft the same motifs? Or does this point to a lost, prehistoric exchange among skywatchers—a shared language of the heavens now eroded by time?

This convergence takes on a startling concreteness at Göbekli Tepe, in modern-day Turkey, dated to around 9600 BCE. This site has unsettled long-held assumptions about when—and how—complex symbolic thought emerged. Its monumental rings of stone, carved with animals, abstract signs, and possible celestial symbols, have been read by some as more than shrines: as astronomical records. One especially tantalising theory suggests that a particular enclosure captures the arrangement of the sky during the Younger Dryas, perhaps preserving the cultural memory of a comet strike that altered climate and myth alike. While contentious, such interpretations raise an arresting possibility—that our earliest monuments were not inspired by farming or political power, but by cosmic unease: the human compulsion to remember, to warn.

Yet, not all sky stories were born of meticulous celestial accounting or catastrophic encounters. Many crystallised from humanity's raw encounter with awe and uncertainty. Comets, with their sudden, unpredictable brilliance, were perfect candidates for divine messengers. In Aztec cosmology, they presaged disaster. Ancient Chinese chronicles, stretching back thousands of years, note their arrivals in meticulous detail, with emperors turning to court astrologers for interpretations. Medieval Europe regarded Halley's Comet in turns as God's anger and as a summons to arms. The content varies, but the underlying instinct is consistent: the conviction that the sky is not indifferent—it reacts, it cautions, it remembers.

Today, we know comets as icy wanderers from the Oort Cloud, devoid of omen in any literal sense. Still, the stories endure—and perhaps they should. They contain not just superstition, but a form of pattern recognition. Some modern hypotheses—about events from the Younger Dryas impact to Tunguska—echo in oral traditions long before science confirmed them through craters and ice cores. In such cases, myth is not a veil over reality but a vessel for it, carrying data in symbolic form. The line between folklore and fact is not necessarily a wall; it can be a bridge.

None of this is to claim that every myth conceals a star map or that every alignment hides prophecy. But the persistent correlation between mythic imagery and astronomical phenomena demands closer study. The ancients were not merely lyrical—they were methodical observers. The sky did not only inspire; it instructed. The heliacal rising of Sirius foretold the Nile's flooding; Orion's disappearance helped Greek sailors time their voyages. Yet because these practices came wrapped in ceremony and story, modernity often discards them as superstition. This is a misreading. The story was the interface; the heavens, the operating system.

Even now, cultural astronomers—be they Indigenous elders, hobbyists, or maverick researchers—continue this decoding. Combining oral tradition with modern star-mapping tools has yielded striking confirmations. Polynesian navigation chants, long dismissed as metaphor, trace precise stellar routes used for open-ocean travel without instruments. Hopi and Navajo traditions chart Venus's cycles with a precision that would impress a contemporary ephemeris-maker. These are not curios of folklore; they are compressed repositories of empirical observation, engineered for transmission across generations.

The deeper truth is that astronomy was never divorced from culture. It was born from its very fabric—from ritual, from myth, from the practical demands of survival. The Enlightenment's great misconception was to imagine science as arriving after myth, as though one replaced the other. In reality, myth may have been science's first incarnation: not falsehood, but framework; not literal record, but analogical model—adaptable, responsive, like the stars themselves, drifting yet enduring. Its purpose has always been the same: to synchronise human life with the grand cycles above.

This is why the study of mythic astronomy matters. It threads together archaeology, astrophysics, anthropology, and narrative theory. It asks what we remember, and how we choose to keep that memory alive. And it restores to the night sky what modernity has stripped away—its role as a living archive. The heavens know nothing in the human sense, and yet they carry everything we have asked them to hold. To read them anew, through the voices carved in stone and spoken in story, may be to recover not just lost knowledge, but an older intimacy with the universe—a reminder that we have always looked for meaning not only in the soil beneath our feet, but in the stars above our heads.

Constellations as Proto-Calendars

There is a subtle brilliance in the way the ancients looked at the stars—not as random pinpricks of light, but as reliable instruments for measuring time, pattern, and destiny. What modern eyes see as decoration, earlier civilisations recognised as structure. Long before mechanical clocks, sun dials, or printed calendars, constellations operated as proto-chronometers, allowing pre-modern societies to synchronise their agricultural, ritualistic, and migratory rhythms with astronomical cycles. But crucially, these star maps were not merely observed—they were mythologised. In this way, early skywatchers embedded calendrical knowledge in durable, symbolic frameworks: myths.

Myth, in this context, was not a matter of fiction—it was mnemonic infrastructure. It encoded seasonal shifts, solstices, equinoxes, heliacal risings, and celestial recurrences into stories that could be memorised, transmitted, and ritualised across generations. The narrative was the medium; the astronomical cycle, the message. Whether it was Orion rising over the Nile to herald the inundation or the periodic disappearance of the Pleiades signalling agricultural or seafaring readiness, the ancients wove their cosmological reality into stories that educated without requiring written literacy.

Orion provides one of the most compelling examples. In ancient Egyptian cosmology, the constellation was identified with Osiris, the god of the afterlife, vegetation, and cyclical renewal. Each year, the heliacal rising of Sirius (closely aligned with Orion) preceded the annual flooding of the Nile, upon which Egyptian agriculture depended. The myth of Osiris—his death, dismemberment, and eventual resurrection—thus becomes more than religious allegory. It serves as an analogue for the river's retreat and return, for the cycle of death and rebirth in nature, and for the interdependence between the sky's rhythms and Earth's fertility. The myth provided both existential reassurance and practical instruction: it told people when to sow, when to mourn, and when to expect the river's life-giving waters.

In the Greek tradition, the same cluster of stars was linked to the hunter Orion—a figure often pursued by or pursuing danger, reflective of seasonal battles between light and darkness, growth and dormancy. The fact that this same formation—the distinctive 'belt' of three stars and its flanking quadrilateral—was mythologised so consistently across cultures suggests it had observational significance long before it had symbolic embellishment. The constellation's visibility during key parts of the year made it an ideal anchor point in the sky, and its stability ensured that myths associated with it would maintain functional accuracy over long timescales.

The Pleiades present an even more intriguing case of cross-cultural calendrical encoding. Known as the Seven Sisters in Greek myth, this star cluster features prominently in legends spanning Japan (*Subaru*), the Maori of New Zealand (*Matariki*), the Cherokee and Lakota of North America, and the Aboriginal Australians. Despite enormous geographic separation, these stories almost always describe the stars as a group of sisters—typically pursued by a male figure (often Orion or a similar figure), often linked to fertility, navigation, or the changing seasons. In many agricultural societies, the heliacal rising of the Pleiades marked the beginning of the planting season. In others, it signified when to begin ocean voyages or prepare for seasonal migrations. That these myths tend to emerge in equatorial and mid-latitude cultures—where the Pleiades are highly visible—speaks to a form of convergent cultural evolution: different groups solving the problem of timekeeping with the same celestial prompt.

It is tempting to dismiss these similarities as coincidences, but this interpretation falls short of the empirical evidence. Modern archaeoastronomy—the interdisciplinary field blending archaeology, astronomy, and anthropology—has uncovered compelling examples of megalithic and early settlement sites that appear to be astronomically aligned. From the 5,000-year-old passage tomb at Newgrange in Ireland (which illuminates during the winter solstice) to the Nabta Playa stone circle in Egypt's Western Desert (arguably predating Stonehenge and possibly aligned with Sirius), these structures suggest that early humans invested enormous effort into recording celestial events in physical form. But when stone was unavailable or social mobility required portability, myths became the more flexible alternative. They served as intangible observatories—encoded in memory, reiterated in ritual, and maintained through storytelling.

This suggests an important reframing: the myth was not merely a cultural product—it was a data container. Consider the alignment of temples and pyramids in Mesoamerica. The Pyramid of Kukulkan at Chichén Itzá, built by the Maya, casts a serpent-shaped shadow along its balustrade during equinoxes. To a casual observer, this may seem like aesthetic coincidence. But within the broader context of Mayan myth—where Kukulkan, the feathered serpent, embodies both celestial and earthly forces—it becomes a synchronisation point: myth meets solar precision. The builders encoded astronomical data into architecture, but also into narrative. The same is true of the Inca, whose ceque system of sacred pathways radiating from Cuzco also mapped solstitial alignments and stellar risings.

The myth-as-calendar hypothesis gains further strength when considered through the lens of information theory. Oral traditions, especially in pre-literate societies, had to maintain fidelity across centuries. Redundancy, rhythm, and symbolic vividness made stories more memorable and resistant to corruption. Embedding calendrical functions into myth allowed information to survive environmental disasters, regime changes, and cultural diffusion. It also enabled societies to maintain synchrony with agricultural cycles and ritual calendars without relying on written records. This was not primitive guesswork—it was high-functioning resilience.

Of course, not every myth can be reduced to an astronomical code, nor should we adopt a hyper-reductive stance that explains away spiritual or psychological meaning. But what is striking is how often ritual calendars—festivals, planting dates, rites of passage—coincide with solar, lunar, or stellar markers. From the Eleusinian Mysteries in ancient Greece to the Dogon's Sirius-based rituals in Mali, the celestial underpinnings of earthly activity were not incidental. They were foundational. And the more we examine these connections, the more we see myth not as falsehood but as *structured metaphor*, anchored in empirical observation.

Modern computational tools have made it possible to reverse-engineer many of these correlations. By using planetarium software to rewind the sky thousands of years, researchers can test whether mythic events correspond with unique astronomical phenomena—eclipses, conjunctions, supernovae, or comet flybys. One intriguing example is the work of researcher Hamlet Zurabyan, who correlates specific star configurations with narrative sequences in the Armenian epic cycle. Similarly, recent studies of the Norse myth of Ragnarok—featuring darkness, fire from the sky, and world-destroying floods—have been reanalysed as potential mythic memories of volcanic eruptions and impact events.

But it is not only ancient cultures that encode sky into story. Even contemporary religious narratives continue this tradition, albeit less explicitly. Consider the Magi following a "star in the east" in the Gospel of Matthew. Scholars have debated for centuries whether this refers to a planetary conjunction, a comet, or a symbolic motif. Yet the impulse remains the same: to frame significant earthly moments within a celestial context. The heavens validate and anticipate the sacred on Earth.

Ultimately, the constellations are not merely decorations of the night sky, nor are myths mere entertainment. They are intertwined systems—one of physical regularity, the other of cultural interpretation. What the ancients achieved, without the benefit of modern instruments, was a kind of cognitive astronomy: a practice of weaving together observation, meaning, and social rhythm into enduring narratives. Those stories were not random. They were chosen. They were repeated because they worked—ritually, agriculturally, cosmologically.

To understand them is not to demystify them. It is to see more clearly how myth and science, rather than being opposed, have long been allies in the human attempt to organise time, tame uncertainty, and align our fragile lives with something that feels permanent. The stars may not speak, but we have always listened. And in listening, we created the first calendars—not on paper, but in story, and not by formula, but by constellation.

Gobekli Tepe and the Zodiac

There are few archaeological sites that confront our assumptions about prehistory more forcefully than Göbekli Tepe. Located in the arid hills of southeastern Turkey and dated to around 9600 BCE, it predates Stonehenge by six millennia and the Great Pyramid by at least seven. The site's monumental T-shaped pillars, some over five metres tall and weighing up to 20 tonnes, are not only among the oldest known megaliths on Earth—they also feature an array of carved animals, abstract symbols, and anthropomorphic figures whose meaning remains contested. But the most provocative theory gaining momentum in recent years is this: that Göbekli Tepe is not simply a temple, not merely a ritual centre, but an astronomical device—an early zodiac encoded in stone, designed to track the sky.

To entertain this possibility is to reframe what we think we know about human cognition in the late Pleistocene. According to the standard model, the Neolithic Revolution—that shift from foraging to farming—occurred because environmental conditions stabilised, leading to sedentary life, surplus production, and eventually, symbolic culture. But if Göbekli Tepe was built before the rise of agriculture and already exhibited astronomical sophistication, then cultural complexity preceded domestication—not the other way around. The implications are not just chronological; they are epistemological.

Several researchers have argued that the animal carvings on Göbekli Tepe's pillars correspond to specific constellations. The most cited among them is the team led by Martin B. Sweatman and Dimitrios Tsikritsis, who published a controversial paper in 2017 linking the imagery on Pillar 43—also known as the "Vulture Stone"—to the constellations visible during a hypothesised comet impact around 10,900 BCE, associated with the Younger Dryas cooling event. According to their interpretation, the vulture represents Sagittarius, the scorpion represents Scorpio (a far less controversial link), and other animals correspond to constellations around the galactic centre. They suggest the entire panel functions as a date stamp for a cataclysmic astronomical event preserved in myth.

Critics have accused this reading of being overly selective and retrofitted, but the methodological questions it raises are valid. If multiple distinct animal symbols found across Göbekli Tepe correspond consistently to constellations still recognised today, can we continue to call the association coincidental? More importantly, why would a hunter-gatherer society with no metal tools or wheel technology invest enormous labour into constructing a site that serves no immediately evident practical function—unless the function was temporal, astronomical, cosmological?

The idea of Göbekli Tepe as a celestial observatory gains further credibility when considered in light of ethnographic parallels. Across traditional cultures, especially those with strong oral traditions, celestial cycles are not abstract—they're vital. They govern planting and harvest, ritual timing, animal migrations, and seasonal migrations. The sky was a stable referent in a world full of flux. For a society at the edge of the Mesolithic–Neolithic transition, predicting environmental rhythms could have meant the difference between feast and famine. Knowledge of the stars wasn't ornamental—it was existential.

Göbekli Tepe's central pillars, often arranged in circular enclosures with peripheral pillars facing inward, suggest not just architectural sophistication but possibly intentional orientation. Some enclosures align with significant solar events—particularly solstices—though the jury is still out on whether this was by design or incidental. However, even the sceptics admit that the builders were doing more than piling stones for shelter. These were sanctuaries, carefully laid out, with iconography far too deliberate to be decorative alone.

If we accept the possibility of Göbekli Tepe as a proto-zodiac, the philosophical implications are vast. It would imply that symbolic abstraction—the capacity to represent time, fate, or divinity in spatialised and narrative form—existed not as a by-product of agrarian sedentism, but as its potential precursor. This flips the chronology of cognitive archaeology on its head. It also forces us to consider the psychological function of myth in a new light. Rather than being fanciful explanations for natural phenomena, myths may have served as mnemonic scaffolds for celestial data—stories wrapped around sky maps, making complex astronomy memorable across generations.

The iconography itself invites layered interpretation. The recurring presence of foxes, for example, could indicate both terrestrial significance (as prey or predator) and celestial reference (perhaps Canis Minor or another now-forgotten constellation). The headless anthropomorphic figures carved into several pillars evoke the mythological decapitations common in flood myths and cosmic catastrophes. Could this be an allegorical record of a disaster, seen in the sky and remembered in stone?

One of the challenges of interpreting Göbekli Tepe lies in its uniqueness. No written language existed to clarify its meanings; no direct descendants preserved its rituals in textual form. This opacity has led to wildly divergent interpretations—from astronomical coding to shamanic ceremonialism, from animistic totemism to social hierarchy cementation. Yet the very multiplicity of these theories suggests a richness of function. Perhaps Göbekli Tepe was not one thing, but many: a calendrical tool, a myth-making engine, a funerary site, a temple to the sky.

The site's deliberate burial around 8000 BCE adds another layer of intrigue. It was not destroyed by enemies or weathered by time—it was intentionally entombed by its own makers. Why? Some scholars speculate that the end of its use corresponded with shifts in subsistence strategies; others suggest a cosmological reason, that once its calendrical or ritual functions had outlived their usefulness—or been fulfilled—it was sealed. The burial might have been a kind of ritual punctuation, a resetting of the world's symbolic logic.

The notion of a "zodiac of stone" also invites reconsideration of the origins of astrology—not as superstition, but as structure. In its earliest forms, astrology was not concerned with individual personality traits or daily horoscopes, but with the harmonics of time. It tracked the heavens to anticipate patterns—weather, war, fertility, disaster. The line between astronomy and astrology was not yet drawn; both were attempts to model regularities in an irregular world. Göbekli Tepe, then, may represent one of the earliest physical manifestations of that ambition: an attempt to make the invisible visible, to render cosmic cycles tangible, to inscribe the heavens into stone.

Scepticism remains essential. The dangers of over-interpretation—projecting modern astronomical categories onto ancient minds—are real. But the impulse to dismiss pattern entirely, to assume that the builders were incapable of intentional alignment or symbolic reference, risks underestimating the cognitive reach of early Homo sapiens. If Homo sapiens was anatomically modern by 200,000 years ago, and behaviourally modern by at least 50,000 years ago, then by 9600 BCE, symbolic cognition was not a novelty—it was a toolkit. Göbekli Tepe might be the oldest surviving artefact of that toolkit applied at monumental scale.

This is not to say the zodiac we know today—Aries through Pisces—was fully formed at Göbekli Tepe. But the impulse to divide the sky into symbolic sectors, to track the movement of celestial bodies through narrative and metaphor, appears to have been underway. That process didn't begin in Babylon or Egypt; it may have far deeper roots, buried beneath millennia of sediment and ideological sedimentation.

To read Göbekli Tepe as a star map is to accept that intelligence leaves traces not only in tools but in stories. It suggests that the sky was not a passive background but an active protagonist in the human drama—watched, charted, revered, and reimagined. And it implies that our desire to find meaning above us may be among the most ancient traits we possess—not a superstition to be discarded, but a heuristic to be decoded. The sky was never silent. Göbekli Tepe tells us we were listening far earlier than we thought

Comets as Divine Messengers Investigates

Few celestial phenomena have inspired as much awe, dread, and symbolic potency as comets. Their unpredictable appearances—seemingly violating the ordered choreography of the night sky—

marked them as omens, intrusions, or even divine agents. Across civilisations and epochs, comets were rarely seen as neutral. They blazed through the firmament like cosmic punctuation, often signalling rupture, reckoning, or revelation. From the apocalyptic imagery of the Book of Revelation to the fatalistic auguries of the Aztec codices, comet sightings have been consistently framed not as mere astronomical curiosities, but as messengers of divine or catastrophic intent. That this pattern recurs globally, and often with striking narrative parallels, demands more than casual explanation.

The ancient sky was not just a canvas—it was a scripture. For early skywatchers, the heavens were alive with meaning. Stars moved with reliable precision, planets traced elegant paths, and eclipses, though dramatic, were cyclical and therefore ultimately decipherable. Comets, by contrast, arrived unannounced. They burned with erratic motion, tailed with unearthly light, and then vanished without pattern. This violation of celestial regularity invited mythic framing. In a cosmos that was otherwise ordered, comets suggested divine intervention—often punishment. The earliest written references to comets, found in Mesopotamian astronomical tablets, already reflect this framing. The scribe did not merely record appearance; they interpreted consequence.

Take the appearance of Halley's Comet in 66 CE. Roman historians recorded it as a portent of calamity, coinciding with the First Jewish-Roman War and the destruction of the Second Temple in Jerusalem. To Jewish communities interpreting scripture through eschatological lenses, the celestial event echoed prophetic warning. By the time of Revelation's composition decades later, cometary imagery had infused the apocalyptic imagination: "A great star, blazing like a torch, fell from the sky... the name of the star is Wormwood." While scholars debate whether this passage refers explicitly to a comet or metaphorically to a divine agent, the visual parallels are unmistakable. A flaming object falling from heaven and poisoning the waters resonates not only with scriptural language but with the global pattern of comet-as-harbinger.

Mesoamerican civilisations took such symbolism to another level of systematisation. The Aztecs, whose priest-astronomers maintained intricate skywatching calendars, treated comets as potent omens tied directly to political and environmental upheaval. The Florentine Codex, compiled in the 16th century, records how a "smoking star" in the sky was interpreted as an omen of the fall of the empire—an interpretation tragically borne out by the arrival of Spanish conquistadors. Even before Cortés set foot on the continent, the appearance of a comet had primed the collective psyche for collapse. The convergence of celestial sign and terrestrial crisis forged an interpretive model: *the heavens do not merely reflect events—they anticipate them.*

The Chinese record is perhaps the most meticulous. Imperial court astronomers documented comet appearances with remarkable precision over two millennia, not merely for astronomical cataloguing, but because the emperor's legitimacy was thought to be written in the stars. A comet could signify moral decay in the court, unrest among the people, or the need for ritual correction. In the *Book of Han*, a comet seen in 12 BCE is interpreted as a sign that the Emperor had violated ritual propriety. This cultural pattern—where cosmology and governance were tightly linked—was not unique to China, but few traditions preserved it with such bureaucratic exactitude. Here, the comet became not a myth, but a policy concern.

From a psychological standpoint, the role of comets in myth and prophecy can be partially explained by what cognitive scientists call the salience heuristic: rare, dramatic, and emotionally charged events are more likely to be remembered and mythologised. A bright comet stretching across the night sky, especially if followed by famine, plague, or war, naturally invites causal linkage. But salience alone does not explain the consistency of the comet-myth pattern across cultures with no direct contact. There is something archetypal at work—some deep template in the human mind that codes celestial irregularity as divine signal.

This has led some researchers to explore whether real cometary impacts or near-misses seeded the cultural memory that evolved into myth. The Younger Dryas Impact Hypothesis, still contested in the academic mainstream, proposes that fragments of a large comet struck Earth around 12,800 years ago, triggering abrupt climate change and massive extinctions in North America. While the geological evidence remains debated, proponents argue that the impact could have birthed flood myths, sky-fire legends, and even the impetus for monumental building—such as the theorised astronomical alignments at Göbekli Tepe. Whether or not the impact occurred as described, the hypothesis illustrates how scientific and mythic narratives might interleave. A cosmic event need not be apocalyptic in scale to embed itself in cultural memory. A single dramatic sighting, if aligned with crisis, could encode itself in the mythic imagination for millennia.

More recently, scholars such as Victor Clube and Bill Napier have suggested that the repeated appearance of Taurid meteors, possibly linked to a decaying comet, may explain why ancient cultures developed skywatching rituals precisely around the times these swarms were most visible. The timing of festivals, sacrifices, or fire rituals may not have been purely agricultural or seasonal, but partly astronomical—tracking the risk of celestial fire. The link between comets and fire is itself a persistent mythic theme. Comets "breathe fire" in textual descriptions; they are "fiery swords" across the sky; they are "serpents of light" whose tails burn through the night. This linguistic pattern reinforces the idea that what was seen in the heavens came to shape how danger, judgment, and power were conceptualised on Earth.

Perhaps the most enduring feature of comet mythology is its ambiguity. A comet could signify doom—or deliverance. In Norse mythology, the world's end, Ragnarök, is heralded by flaming stars and skies in tumult. Yet in Hindu cosmology, comets (dhumketu) are sometimes considered messengers of transition, the thresholds between yuga cycles. In some Maori traditions, a comet marked the death of a great chief, guiding his soul into the heavens. The motif remains stable: a disruption in the celestial order marks a rupture in the terrestrial one. But the moral reading—punishment or purification—varies by culture, context, and need.

What is striking, then, is not simply that cultures saw omens in comets, but that they *used* those omens to scaffold meaning in times of uncertainty. A comet provided a kind of narrative architecture—linking the cosmic with the communal, the celestial with the psychological. When societies faced destabilisation—whether through war, famine, plague, or succession—a comet's arrival allowed that disruption to be situated within a larger, often divine, schema. It wasn't just an explanation; it was a theodicy.

Even in modernity, traces of this interpretive reflex persist. Halley's Comet's return in 1910 caused panic across Europe and North America, with newspapers warning that Earth's passage through its tail might "poison the atmosphere." Entrepreneurs sold "comet pills" to protect against imagined toxins. Though the scientific community dismissed these fears, the cultural reaction underscored an enduring psychological truth: we may understand comets better now, but we are not fully immune to their symbolic gravity.

Contemporary space science has rendered comets less mysterious but no less marvellous. We now know they are icy relics from the solar system's formation, often harbouring the very molecules thought to seed life. The Rosetta mission's landing on Comet 67P in 2014 was a technological marvel, but it also evoked something older—an ancestral awe at touching the messengers of the gods. That we have gone from fearing comets to landing spacecraft on them in less than a century is a triumph of human ingenuity. Yet even now, popular culture—from disaster films to apocalyptic thrillers—casts comets as harbingers of judgment, both feared and revered.

The enduring mythological charge of comets may ultimately rest on a paradox. They are among the most ancient of celestial phenomena, yet they always arrive as if new. Their suddenness defies routine; their beauty cloaks terror. They are messengers whose message is never entirely deciphered. And perhaps that is why they persist—not because we misunderstood them, but because they articulate something deeper: that even in an age of precision and prediction, we remain susceptible to awe. And awe, as myth has long understood, is never neutral. It compels us to search for meaning, even—or especially—when none is promised.

Modern Skywatchers and Ancient Eyes

A few thousand years ago, when the night sky was the only available screen, the act of stargazing wasn't a hobby—it was survival, governance, myth-making, and memory. Today, as telescopes proliferate and astrophotography is accessible from a smartphone, modern skywatchers continue to look upward, but the nature of their gaze has changed. Where ancient observers embedded meaning in myth and ritual, today's amateur astronomers parse the heavens with software, star charts, and satellite feeds. Yet a growing number of them are beginning to recognise a striking continuity: *our ancestors weren't just looking at stars—they were encoding knowledge into stories that still shimmer with astronomical precision.* This convergence of myth and science, of ancient eyes and modern lenses, is beginning to reshape how we understand the cosmos—not just as a physical system, but as a long-standing human archive of orientation, prediction, and meaning.

The divide between astronomy and mythology has long been overstated. For centuries, scientific progress was premised on distancing itself from mythic interpretation. Yet many of the sky's most familiar names—Orion, Andromeda, Perseus—are drawn from classical myth. More importantly, the motifs those names anchor often reflect actual celestial behaviours. Orion's seasonal emergence and disappearance, for example, maps closely to agricultural cycles and ritual timings in multiple ancient cultures. The Pleiades, visible from every inhabited continent, were often seen as a cluster of sisters, birds, or mourners—nearly always female figures tied to fertility, timekeeping, or harvest. These resemblances are not coincidental. *When oral tradition intersects with regular celestial patterns, we must ask: what is being remembered, and how precisely?*

Software now allows amateur skywatchers to simulate the sky from any place on Earth at any point in history. Tools like Stellarium, SkySafari, or Starry Night aren't just aids for astrophotography—they're retrospective lenses into the mythic imagination. By aligning archaeoastronomical data with myths, users can track when certain constellations rose at dawn or marked solstices in a given region. This has led to a renaissance in decoding ancient sites—from Stonehenge to Chaco Canyon, from the Maltese temples to the megaliths of Nabta Playa—as astronomical instruments wrapped in symbolic form. Gobekli Tepe, often cited as the oldest known megalithic structure, is now at the heart of debates not only about pre-agricultural cosmology but about whether certain animal carvings map star positions during specific epochs. It is a debate no longer confined to academic journals—forums, podcasts, and skywatching communities are actively joining the interpretive process.

What makes this convergence especially valuable is not only the recovery of knowledge but the rebalancing of authority. Indigenous skywatching traditions—long dismissed as poetic or primitive—are gaining credibility among astronomers and anthropologists willing to question the assumption that scientific understanding must be recent or Western. The Yolngu people of northern Australia have known for centuries that the moon's phases affect tides—a fact Western science only formalised in the 17th century. In the Arctic, Inuit oral histories describe the sun's seasonal shifts with linguistic precision that rivals GPS alignment. Polynesian navigators used star compasses and horizon cues to cross thousands of kilometres of open ocean without modern instruments. *These were not mythic beliefs—they were sophisticated systems of empirical observation encoded in story, chant, and symbol.*

As interest in "ethnoastronomy" grows, modern amateur astronomers increasingly find themselves not just peering through telescopes but also poring over mythology, folklore, and ancient texts. In doing so, a fascinating epistemic shift occurs: the night sky becomes not a blank canvas to be interpreted in purely empirical terms, but a palimpsest layered with ancestral insight. Recognising this does not require abandoning scientific rigour—it requires expanding the definition of what counts as data. A myth that aligns with the heliacal rising of Sirius across millennia is not a coincidence—it is a calibration tool. Oral memory, when stabilised across generations, becomes a kind of *temporal software*, capable of preserving astronomical cycles well before writing systems emerged.

Take the Dogon of Mali, often cited—controversially—for their apparent knowledge of Sirius B, a white dwarf companion to Sirius A, invisible to the naked eye. While some claims around their cosmology remain speculative or exaggerated, the case has pushed scholars and skywatchers alike to reconsider whether traditional knowledge systems might encode observational truths in metaphoric dress. Likewise, the Navajo story of the "Revolving Male" and

"Revolving Female" stars—linked to circumpolar constellations—may be a way of preserving sky rotation knowledge across seasonal cycles. These are not curiosities; they are alternative frameworks for organising astronomical understanding.

For the modern skywatcher, integrating these perspectives opens several transformative pathways. First, it dissolves the false dichotomy between myth and measurement. A story that encodes a lunar eclipse cycle is not less scientific because it uses gods or animals as mnemonic anchors. Second, it reveals astronomy as a fundamentally human endeavour—one that has always been situated within a cultural, spiritual, and ecological matrix. Third, and perhaps most radically, it invites the possibility that *our ancestors were not merely imagining the cosmos—they were quantifying it through narrative form.*

This realisation has implications for how we teach, research, and practice astronomy today. Rather than segregate myth into the humanities and astronomy into the sciences, interdisciplinary models are emerging that treat ancient celestial stories as datasets in their own right. Initiatives like the Indigenous Sky Stories project in Australia, or the University of California's cross-disciplinary archaeoastronomy programmes, are equipping a new generation of scholars and citizen scientists to read the sky both empirically and culturally. In these settings, amateur astronomers are not merely observers—they are translators, bridging the technical and the mythic, the modern and the ancestral.

Even the language of technology is adapting. "Sky culture" software layers constellations according to multiple traditions—Western, Chinese, Lakota, Greek, Inca—allowing users to toggle between cosmologies. This multiplicity resists the colonial tendency to flatten all stargazing into one narrative. Instead, it reveals *a universe that was always plural*, perceived through many lenses, yet anchored in shared human wonder. The fact that so many traditions locate the Pleiades as a cluster of sisters or the Orion belt as a warrior figure suggests not just archetypal projection but shared cognitive strategies for mapping celestial regularities.

There is, of course, a risk of romanticising the past or over-interpreting ancient alignments. Not every stone circle encodes an eclipse cycle; not every myth points to a solstice. But the growing dialogue between modern skywatchers and ancient symbolic systems invites a critical humility. It forces us to ask: what did we lose when we separated measurement from meaning? And what might we regain by reintegrating them?

Perhaps the most urgent relevance of this dialogue lies not in antiquarian interest, but in environmental and existential awareness. Ancient skywatchers did not separate the sky from the land, or the calendar from the body. When Orion rose, it was time to plant; when the moon waned, certain rituals were forbidden. The sky was not just observed—it was lived. In a time when light pollution, climate crisis, and digital abstraction threaten our planetary orientation, revisiting these older sky-cultures is not nostalgic—it is necessary. *To recover a living relationship with the stars may be to recover a sense of scale, of rhythm, of ecological embeddedness that modern life has all but erased.*

In the end, what unites the modern astrophotographer adjusting a telescope app with the tribal elder recounting a sky-story is a shared act: looking up with intention. That act—ancient and ever-new—is where meaning begins. It is not merely observation, but participation in a dialogue as old as consciousness. When modern skywatchers learn to see not just stars, but *stories as coordinates*, they join a lineage far older than any scientific discipline. They become, in the truest sense, readers of the sky. And perhaps that is what our ancestors always intended: not just to impress their myths upon the heavens, but to inscribe a method—one that still works, if we remember how to read it.

Chapter 5
Beasts of Mind and Soil

Cryptids have, for the most part, been relegated to the dusty margins of serious inquiry—a sort of cultural attic filled with curiosities that respectable science prefers not to touch. The very word calls up images of grainy photographs and dubious late-night documentaries: the hulking outline of a retreating Sasquatch, the ripples of something large in Loch Ness, or breathless accounts of winged humanoids glimpsed in some remote ravine. Yet, under the tabloid sheen lies a stubbornly neglected question: why, across centuries and continents, do human beings keep imagining the same kinds of creatures? And more intriguingly—what ancient truths might those inventions be sheltering, as though in the fossil record of the mind? When we listen to tales of the Wild Man, the forest ape, or the predator whose eyes flash in the dark, perhaps we are not witnessing mere fantasy. Perhaps we are brushing up against evolutionary memory, disguised as myth.

This chapter suggests turning the usual assumption inside out. Instead of thinking myths create monsters from nothing, what if they distil the lingering psychological residue of once-real beings, behaviours, or threats? In the growing convergence of palaeogenetics, archaeology, and ethnography, the neat border between "imagined" and "extinct" grows harder to defend. The discoveries of archaic humans—Homo floresiensis in Indonesia, or the shadowy Denisovans revealed through scraps of bone and strands of DNA—have forced more than one uncomfortable revision. Tellingly, these species were not first known to science through fossil beds, but through stories: islanders spoke of the Ebu Gogo, a small, food-stealing, speech-mimicking humanoid. It was dismissed as island folklore until, in 2003, bones surfaced that matched the legend uncannily. What science had long called superstition now occupies a place in the catalogue of prehistory.

Not every fireside story conceals a misclassified species, of course. Yet some recurring archetypes—the hairy biped, the part-human hybrid, the soul-hungry predator—are too consistent in detail and too deeply resonant across disparate cultures to be mere coincidence. These figures appear in Amazonian chants and Himalayan epics alike, hinting at what might be called evolutionary hauntings. They may carry the imprint not of individual creatures, but of whole categories—species-types, ancient rivals, uneasy neighbours in the human story. For much of our deep past, we shared the planet with other intelligent hominins. We competed, interbred, learned from—and evaded—them. That truth is only just regaining its place in the scientific mainstream. Myths, apparently, never let it go.

Consider the Wild Man. He turns up everywhere: Sasquatch in North America, Yeti in Tibet, Yowie in Australia, Almasty in Central Asia, Orang Pendek in Sumatra. This is not a case of interchangeable boogeymen. Across vast distances, they retain remarkably similar traits: upright stance, preference for forest or mountain refuges, nocturnal foraging, vocal mimicry, territorial bluster, and rudimentary communication. Were these simply random inventions, we'd expect far more variation. Instead, we find structural repetition—suggesting the brain might store not only personal memories, but intergenerational warnings coded into story and taboo. When a child is told to avoid the forest lest the Wild Man take them, the warning works on two levels: it disciplines behaviour here and now, and it resonates with an ancient awareness that our species once shared those woods with something that was not quite human.

Neuroscience is beginning to catch up with folklore. Emerging research on inherited fear circuitry proposes that certain trauma-based memories can be passed on epigenetically as well as culturally. This could explain the uncanny potency of certain predator archetypes in dream, religion, and oral history. You do not need to have seen a predator for its image to raise your pulse; symbols can awaken ancient reflexes. Take the chimera—those spliced-together beasts of myth. None exist in their exact described form, yet hybridisation is no stranger to evolution. We now carry genetic traces of Neanderthals and Denisovans, after all. Perhaps the human mind preserved such mingling in metaphor: part-man, part-beast, not as an invention of fantasy but as a symbolic wrestling with blurred species boundaries.

Chimeras often inhabit the realm of boundary violation. From the sphinx of Egypt to Persia's griffin, to Mesoamerica's feathered serpent, they bring together attributes that strain categorisation: flight and ferocity, cunning and savagery. They confound the mind's taxonomies—human versus animal, divine versus monstrous, self versus other. It is no accident they so often guard thresholds—temple gates, sacred springs, cave mouths. These liminal zones are places of change and danger. To place a hybrid guardian there is, in effect, to signal the brain's own alarm: here the categories break down; tread carefully.

If hybrids mark our mental caution zones, then purely predatory cryptids—the spectral black dog, the thunderbird, the chupacabra, the flying serpent—are perhaps condensed emblems of trauma. Evolution wired our perception for speed when it comes to certain dangers: snakes, talons, sudden aerial descents, glowing eyes in the dark. Many cryptids exaggerate precisely those features. These images keep returning, not because they are irrational, but because they once mapped directly onto life-and-death realities.

This chapter will not try to prove that such creatures walk the earth today. That is a different, often theatrical, pursuit. Our concern is what these beings reveal about us—how myths safeguard cognitive instincts, how stories encode the thresholds of evolution, and how the mind retains ghostly traces of long-lost neighbours. We will draw from fossil finds like Homo luzonensis and Homo naledi, asking what their presence in our family tree means for the mental landscapes of early humans. And we will entertain the possibility that some myths are not inventions at all, but faded recollections of genuine encounters.

Beyond anthropology or neurology lies a philosophical question: what does it say about us that we still dream of predators gone for thousands of years? Why do children, oceans apart, draw the same red-eyed shadows with claws? Why does the Wild Man persist, unbanished, even after every rational dismissal? What cryptids truly preserve is not zoology, but memory of the boundaries we once tested—the frontiers we crossed, the intelligences we met, the dangers we outran.

In the end, "cryptid" may be the wrong label. It implies something hidden, waiting to be discovered out there. Perhaps these beings were never hidden at all. They live, instead, in the winding corridors of human memory, carried forward by the species that survived them. Myths, in this light, are not lies—they are archives for what the fossil record cannot hold: subjective, embodied, encrypted knowledge from an age when survival meant constant negotiation with other minds. Our task here is to excavate those archives—not to take their creatures as literal truth, but to recognise them as memory-systems, shaped by survival, sharpened by threat, and haunted by kinship.

Sasquatch, Yeti, and the Wild Man Archetype Beyond hoaxes:

Sasquatch. Yeti. Almasty. Orang Pendek. Names vary by region, but the silhouette remains constant: a large, bipedal, hair-covered humanoid, elusive yet strangely familiar. Mainstream discourse usually sorts such figures into two categories—either hoaxes perpetuated by cranks or folkloric holdovers from less enlightened times. This false binary misses the most important question: *why does this archetype appear independently in so many cultures, with striking consistency, despite no apparent contact between them?* What is the Wild Man a memory *of?*

To treat the Wild Man as pure fantasy is to overlook the deeper substrate on which these myths rest—one that overlaps with both palaeoanthropology and evolutionary psychology. The Wild Man, unlike other mythological figures, is not fantastical in form. He is not a shapeshifter, winged beast, or spectral deity. He is merely *almost human*—larger, hairier, silent, reclusive, possibly intelligent, and invariably liminal. He dwells at the edge of the forest, the foot of the mountain, the mouth of the cave. In this, the Wild Man may function not as an invention but as a *residue*—a cognitive imprint of a time when Homo sapiens was not alone.

Archaeological evidence now confirms that until relatively recently—on an evolutionary timescale—we coexisted with multiple hominin species. *Homo floresiensis*, known colloquially as the "hobbit," stood barely a metre tall and persisted on the Indonesian island of

Flores until at least 50,000 years ago. Nearby islanders had long told stories of small, hairy people called the Ebu Gogo, said to steal food, mimic speech, and flee from fire. These reports were widely dismissed until the fossil record proved them uncannily predictive. Similarly, genetic evidence from Denisovan DNA, found in modern Melanesian and East Asian populations, suggests that our ancestors interbred with a ghost population—one never physically discovered, but indirectly revealed through molecular remnants. *Folklore, it seems, sometimes outruns the spade of archaeology.*

This raises a disquieting possibility: that myths of the Wild Man might be degraded memories—not of one species, but of a *type*. The human mind, shaped under the long shadow of interspecies rivalry and coexistence, may have encoded these encounters in narrative form. Such myths were not ornamental—they were survival algorithms, ancestral warnings, and perhaps even distorted histories of genetic contact. The consistent tropes—bipedal gait, forest habitat, mimicry of language, nocturnality—could signal behavioural truths abstracted into story. If so, the Wild Man is not fictional. He is an evolutionary echo.

Consider the case of the Yeti in Himalayan cultures. Far from being a fringe belief, the creature appears in monastic texts, village lore, and ritual practices. In Tibetan Buddhism, the Mi-Go (another name for the Yeti) is treated not merely as a beast, but as a morally ambiguous presence—sometimes a threat, sometimes a guardian of sacred places. Local knowledge speaks of specific migratory paths, vocalisations, and reproductive behaviours. This level of ethnographic specificity challenges the Western caricature of the Yeti as a tabloid curiosity. It suggests that Himalayan communities have preserved a structured, functional taxonomy of the Wild Man—one not easily dismissed as irrational.

North American First Nations also maintain longstanding beliefs in the Sasquatch, a term derived from the Salish word *Sésquac*, meaning "wild man of the woods." These accounts predate European colonisation and vary little over time. The creature is described as territorial but not malevolent, capable of complex communication, and associated with specific ecological zones. Some tribal elders interpret the Sasquatch not as an animal but as a *people*, albeit a different kind—more attuned to the natural order, less corrupted by technological arrogance. This perception invites ethical reflection: what if the Wild Man is not a lesser being, but a *parallel lineage*, extinguished not by failure but by proximity to Homo sapiens?

The evolutionary implications are sobering. Recent fossil finds—such as *Homo naledi* in South Africa—demonstrate that small-brained hominins persisted far longer than previously thought, with some possibly alive within the last 300,000 years. We also know that tool use, symbolic marking, and complex social behaviour were not unique to Homo sapiens. Neanderthals buried their dead. Denisovans left rings and beads. The boundary between "us" and "them" grows blurrier by the year. If a remnant population of archaic humans lingered in remote environments, particularly in ecologically stable refugia such as the Himalayas, Pacific Northwest, or Southeast Asian islands, our only record might be myth.

The Wild Man archetype also plays a critical role in cultural psychology. He marks the boundary between civilisation and wilderness—between the known and the unknowable. Unlike wolves, bears, or other apex predators, he carries an uncanny quality: he is *watching*. He is intelligent but silent, visible but unreachable. This evokes a unique kind of fear—not of being eaten, but of being *observed* by something that might once have been kin. The Wild Man thus becomes a mirror—not of what we are, but of what we might have been under different evolutionary pressures. He embodies the path not taken.

There is also a strong psychoanalytical reading. The Wild Man operates as a projection of our own suppressed instincts—the part of the human psyche that refuses domestication. In European medieval art and myth, the Wodewose—literally the "wood-being"—symbolised the chaotic, pre-social aspect of the self. Often covered in hair, wielding a club, and unbound by law or decorum, he was feared and fetishised. During periods of social repression, particularly in the late Middle Ages, the Wodewose became both scapegoat and fantasy—a vessel for all that had been exiled from polite society. Similar dynamics appear in the Japanese "Yamabito," the mountain people said to descend from lost tribes, hiding from modernity in remote valleys.

Across cultures, then, the Wild Man serves dual functions: as evolutionary memory and as symbolic container. He holds our ancestral experience of cohabiting with other hominins, and he expresses our cultural anxiety about the collapse of boundaries—between human and animal, civilised and wild, known and hidden. The refusal of this archetype to disappear suggests not that he is "out there" in the woods, but that he is *in here*, embedded in the cognitive architecture of our species.

Scientific disciplines have only recently become willing to entertain this possibility. Interdisciplinary research in evolutionary psychology, cultural anthropology, and genetic archaeology points toward a reappraisal of myth as data—*encrypted knowledge passed through the filter of narrative.* Myths of the Wild Man, when viewed through this lens, are neither lies nor hallucinations. They are fossilised memories: behavioural traits and encounters stored in story form, selected not for empirical accuracy, but for transmissibility and survival relevance.

This reframing has implications for how we treat indigenous knowledge systems. Where once local myths were dismissed as primitive superstition, we now begin to recognise that oral traditions may preserve memory traces far older than writing. In some cases, these traditions have correctly predicted natural phenomena (such as volcanic eruptions or astronomical cycles) long before scientific verification. It is therefore not unthinkable that stories of bipedal hominids, preserved with astonishing consistency across continents, may also contain truths that science is only now rediscovering.

To ask whether Sasquatch or the Yeti "exists" is to ask the wrong question. What matters more is *why* these figures persist, with such tenacity, across time, geography, and cultural transformation. Whether or not we ever capture one in the flesh, their psychological presence points to real evolutionary conditions—of coexistence, competition, and perhaps even kinship. These are not myths in the dismissive sense. They are *mnemonic artefacts*, surviving long after the bones of their subjects have vanished into soil and memory.

To dismiss the Wild Man is to silence an entire archive of cognitive history. To examine him seriously is not to embrace fantasy but to respect the integrity of myth as a form of long-term memory—a memory of species once seen, perhaps once spoken to, and then forgotten by the fossil record but remembered in story. This is the hypothesis worth entertaining: not whether these creatures live now, but whether *we have lived with them before.*

The Hobbits of Flores and the Denisovan Ghosts Real discoveries

For most of the twentieth century, the human story was told in increasingly linear terms: from primitive ancestor to modern sapiens, via a handful of waystations like Homo erectus and Neanderthal, all fading cleanly into obsolescence as our own lineage triumphed through superior cognition, language, and social complexity. By the early 1990s, the prevailing view in palaeoanthropology was one of a relatively simple evolutionary tree with Homo sapiens as its lone surviving branch—solitary, inevitable, and unchallenged. That story has not merely been revised. It has been ruptured.

The discoveries of *Homo floresiensis* in 2003 and the Denisovan genome in 2010 shattered the illusion of a tidy evolutionary path. Instead, they revealed a world in which multiple hominin species shared the Earth at the same time—overlapping in space, genetics, and perhaps even culture. These species were not confined to deep prehistory. They were recent. They walked the Earth long after the invention of fire, long after early art and burial rituals. Some, we now know, live on in our DNA. The mythic and the material have collided with a force that demands a reassessment—not just of the fossil record, but of what we've always called folklore.

The case of *Homo floresiensis*—popularly dubbed the "Hobbit"—is illustrative. Discovered in Liang Bua cave on the Indonesian island of Flores, this diminutive hominin stood about one metre tall, with a brain one-third the size of ours. Yet the archaeological context indicates stone tools, controlled fire, and potential hunting of now-extinct megafauna. More striking than their morphology was their chronology: the original finds dated to as recently as 12,000 years ago, though later analysis suggests extinction may have occurred around 50,000 years ago. Either date is breathtaking. In evolutionary terms, it places *Homo floresiensis* well within the lifespan of anatomically modern humans, and within geographic reach of our migration corridors.

Islanders on Flores, notably the Nage people, have long told stories of the Ebu Gogo—a small, hairy, cave-dwelling people said to steal crops, speak in murmurs, and run with surprising speed. These oral traditions were documented by anthropologists decades before any fossil discovery, but were widely dismissed as cultural fictions. After 2003, they acquired new significance. Here, folklore proved predictive, even diagnostic. That raises a radical possibility: that in some cases, local myths may preserve degraded but meaningful memories of extinct species. If such memories endure in oral tradition, how many other "myths" may be shadows of real encounters?

The Denisovan story pushes the boundary even further—because it unfolds not through bones, but through genes. In 2008, researchers excavating Denisova Cave in Siberia found a finger bone unlike any previously recorded. Genetic sequencing revealed something startling: a distinct species of archaic human, not Neanderthal, not sapiens, but something else. Denisovans were later shown to have interbred with both other groups, and their DNA survives in varying degrees across Melanesian, East Asian, and even some South Asian populations.

The Denisovans have left us no artefacts, no language, and no myths of their own—only fragments and alleles. But their influence may stretch further than the fossil record can currently prove. Consider this: high-altitude adaptation in modern Tibetans—a genetic trait that allows efficient oxygen use at extreme elevations—has been traced to a Denisovan gene variant. This is not metaphor or myth. It is molecular inheritance. In a very real sense, people alive today carry functional adaptations acquired from a species we had never seen until the last fifteen years.

This reverses the traditional direction of anthropological authority. For over a century, Western science claimed primacy over myth: separating fact from folklore, bones from legends, science from story. Yet as *Homo floresiensis* and Denisovan DNA have shown, it is possible—perhaps common—for folklore to encode evolutionary truth, awaiting only the right tools to decode it. This does not mean all myths are literal memories. But it forces us to treat them as potentially *informational*, rather than merely expressive.

There is also a broader philosophical tension here. For generations, modern humans have imagined ourselves as the apex species: the sole surviving genus of a once-branching tree. That narrative buttressed assumptions about our uniqueness, our superiority, our right to mastery. But the picture emerging from palaeogenomics, comparative anthropology, and indigenous epistemologies suggests something different: that we are one thread in a once-crowded fabric of intelligent life. And crucially, that our survival may have owed more to luck, geography, and accident than to cognitive supremacy.

This insight echoes faintly but persistently in ancient texts. In Mesopotamian epics, we read of "wild men of the mountain," hairy and uncivilised but strong and cunning. In Southeast Asian traditions, we hear of forest spirits with human features, said to retreat deeper into the wilderness as villages encroach. In Himalayan lore, the Mi-Go or Yeti occupies a nearly identical conceptual space: intelligent but separate, kin yet other. Are these cultural projections? Almost certainly. Are they entirely imagined? That is far less certain.

Anthropologists now speak cautiously of "cryptomemory"—the hypothesis that certain collective memories can be retained across deep time, particularly in societies with strong oral transmission and intimate ecological knowledge. These memories may not preserve names or dates, but they can encode behavioural templates, habitat associations, and cautionary functions. In this view, the Denisovans and Flores Hobbits are not so much "lost species" as *partially remembered ones*—preserved not in textbooks but in stories, warnings, and spiritual frameworks.

What's remarkable is not that science is finally validating some aspects of myth, but that myth preserved its claims *without* validation. For thousands of years, people spoke of small forest dwellers, mountain men, and hybrid beings. These claims were ridiculed, even pathologised. Only with the advent of molecular sequencing, isotope dating, and interdisciplinary analysis have we come to recognise that these traditions may represent forms of *empirical memory*. They are not scientific in method, but they are observational in origin—and resilient in transmission.

The challenge now is to realign the dialogue between scientific and cultural knowledge. Rather than mining folklore for data points to confirm existing theories, we must approach it with the same epistemic humility we bring to fossil sites. Every story is not evidence. But every persistent motif—especially one tied to specific geographies, repeated over generations—deserves interrogation on its own terms. The Hobbits of Flores and the "ghost" of the Denisovans show us that the boundary between legend and fossil record is not fixed. It is porous. And it is moving.

There is also an ethical implication here. If we are to take these species seriously—not just as objects of curiosity, but as once-living members of the human family—we must grapple with the likely reality that our ancestors encountered them, competed with them, perhaps even exterminated them. The mythic status we've assigned to these species often serves a psychological function: to relegate them to the realm of fantasy, thus excusing ourselves from any historical responsibility. But as the evidence accumulates, so does the weight of recognition. We are not the sole inheritors of the human lineage. We are its survivors.

And so the final irony may be this: that in our drive to disprove myth, we have re-entered it. That the stories told in caves and around fires were not fabrications, but frameworks—ways of holding onto memories too strange to preserve in prose. The "Hobbits" and "ghosts" of archaic humanity, once thought the province of fantasy, now force us to ask how many other fragments remain hidden—encoded not in stone or DNA, but in the strange, persistent language of myth. Perhaps, if we learn to listen carefully, more such memories will surface. And with them, a fuller reckoning of what it means to be human.

The Myth of the Chimera

Few mythological motifs are as consistently unsettling—or as psychologically revealing—as that of the chimera: hybrid beings stitched together from mismatched animal parts, defying the natural order. From the lion-goat-serpent creature of Greek lore to the sphinxes of Egypt, nagas of South Asia, and the feathered serpents of Mesoamerica, composite monsters appear across continents and civilisations, often independently. These unnatural amalgams do more than terrify; they provoke a deeper, often subconscious question: *What does it mean when the body itself becomes a site of contradiction?*

At first glance, the answer may seem cultural or symbolic. The classical Greek chimera—typically rendered as a lion's body, goat's head protruding from its back, and a snake for a tail—was interpreted by later commentators as an allegory of chaos, a warning against hybridity, or a symbol of inner conflict. But this tidy symbolic reading doesn't fully explain the universality or visceral intensity of such creatures. Why would cultures separated by thousands of miles and millennia independently imagine monsters that combine biological elements in anatomically impossible ways?

The chimera is not merely a metaphor; it is a psychological boundary marker. It surfaces when categories break down—when what is human, animal, divine, or monstrous begins to blur. In this sense, chimeras are not simply creations of artistic whimsy, but artefacts of epistemic tension. They emerge precisely when societies confront their own limits of classification—when taxonomy, theology, and biology fail to cohere. *The chimera signifies a collapse of the known into the uncanny*, a rupture in the mental architecture we rely upon to stabilise identity and reality.

The question then becomes: is this collapse purely conceptual, or does it preserve something ancestral?

One provocative hypothesis—still highly speculative, but increasingly relevant—suggests that hybrid myths may stem from ancestral memories of biological realities our cognitive models have since buried. Across the late Pleistocene and early Holocene, Homo sapiens did not live in isolation. We coexisted, interbred, and perhaps even competed with multiple other hominin species. The existence of interspecies offspring is now genetic fact: Neanderthal and Denisovan DNA make up measurable percentages of the modern human genome. But what might a partially hybridised offspring have looked like? And how would such a being have been perceived—not as a population, but as an individual?

It is conceivable that isolated examples of such hybridisation gave rise to local taboos, tales, or representations—not necessarily in the literal sense of lions bred with goats, but in symbolic efforts to reconcile traits perceived as anomalous. An individual with unrecognisable features—extreme prognathism, asymmetry, or an unusual gait—might have triggered a kind of representational reflex: the attempt to describe the indescribable through metaphor. *In this model, the chimera is less a fabrication than a cultural exoskeleton around biological anomaly.*

More compelling still is the possibility that the hybrid creature serves as a symbolic memory of *boundary violation.* In every society, bodies represent more than biology; they are moral territories. The chimera violates this deeply embedded sense of bodily order. It conjoins what must remain separate—fur and feather, fang and hoof, serpent and bird. In this light, chimeras may signal deep anxieties about contamination, hybridity, and unnatural fusion—anxieties that mirror real threats from disease vectors, predation, and genetic mutation. When ancient peoples imagined a creature with the wings of an eagle, the hindquarters of a lion, and the face of a man, they were not just inventing fantasy; they were registering psychological alarms around incoherence and unpredictability.

This reading gains further traction when one considers the role of chimeric motifs in ritual and magic. In ancient Egypt, protective deities like Taweret—part hippopotamus, part crocodile, part lioness—guarded pregnant women from evil spirits. These composite forms were not merely decorative; they were understood to *harness the traits of each constituent animal*, creating beings of layered potency. To think of the chimera as mere monstrosity is to miss its role as a magical technology: *a metaphysical composite assembled for specific cultural work*, whether that be protection, terror, or moral instruction.

One cannot ignore the genetic resonances of these myths in the present. With the advent of CRISPR and synthetic biology, we are now creating real-life chimeras—genetically engineered animals with human cells, pigs with human-compatible organs, and human embryos edited for disease resistance. What was once mythical has become mechanical. In 2021, scientists created chimeric embryos containing both human and monkey cells. The stated aim was medical research, but the ethical boundaries are blurry at best. What differentiates today's laboratories from ancient myths is not only technological capacity, but moral ambiguity. The ancients may have feared chimeras as symbols of unnatural merging. We, in contrast, pursue them in search of cure, efficiency, and control.

This temporal inversion—where myth anticipates science—is not accidental. Myth, at its best, functions as a kind of emotional foresight. It gives form to intuitions before facts are fully available. In the case of the chimera, it suggests that human imagination has long intuited the peril and promise of biological manipulation. The motif recurs because it speaks to something unresolved: *a sense that the human form is both sanctified and permeable, both sacred and incomplete.*

Consider the enduring popularity of the Minotaur, the centaur, the griffin, or the harpy. Each combines human and animal in ways that are narratively potent but biologically impossible. Yet these creatures are rarely presented as pure monsters. They are often guardians, exiles, or victims—liminal beings trapped between worlds. This ambiguity is essential. It implies that hybridity itself is not inherently evil, but *dangerous when it defies order without purpose.* Mythical chimeras are not punished for their forms; they are feared for what their forms reveal: the fragility of human categorisation, the permeability of nature, the arbitrariness of boundaries.

From a psychoanalytic perspective, the chimera also functions as a projection of internal contradiction. The lion, serpent, and goat—each with distinct temperaments, habitats, and mythic roles—may represent conflicting drives within the psyche. The fusion of these into one body externalises psychological dissonance: rage, wisdom, lust, fear. The hybrid creature becomes an *externalised composite of inner tension*, offering both a mirror and a warning. In this sense, the chimera is not a monster, but a *map of inner conflict*, rendered in zoological metaphor.

Where does this leave us? With a richer, more layered understanding of why the chimera endures—not just in ancient myth, but in science fiction, genetic ethics, and digital culture. From cybernetic androids to artificial intelligences modelled on human emotion, the hybrid continues to fascinate because it challenges what it means to be whole. The ancient chimera was built from bone and claw; today's is built from code and chromosome. But the psychological substrate is the same: *an irreducible tension between unity and multiplicity, essence and fusion, nature and artifice.*

In a world increasingly defined by synthetic biology, cross-species experimentation, and machine learning, the myth of the chimera no longer belongs to antiquity. It is alive, recalibrated, and reasserting its relevance. Not because we believe in monsters, but because we are building them—carefully, ethically, and often unknowingly. The line between symbolic and real is narrowing. The chimera stares back not only from the past but from our future, asking not "Do I exist?" but "How far will you go to make me real?"

The Underdream of Predators:

What if our most terrifying myths aren't inventions, but inheritances—psychic fossils of past survival crises encoded in culture, memory, and instinct? Beneath the rational scaffold of modern thought, certain images retain a primal voltage: clawed shadows in the woods, glowing eyes beyond the firelight, the uncanny sensation of being watched while alone. These motifs appear across disparate civilisations, separated by language, geography, and belief systems. Yet they converge with remarkable fidelity around the same core template: the lurking predator. Whether named as werewolf, wendigo, rakshasa, or redcap, many legendary creatures—those that haunt dreams more than they entertain—are *not* expressions of creative whimsy. They are encrypted fear-forms, *resonant echoes of evolutionary trauma* passed down through narrative, shaped by selection pressures rather than storytelling trends.

This idea—at once anthropological, neurobiological, and mythographic—positions predators not just as ecological threats but as psychological architects. For hundreds of millennia, predation was the central problem of our survival. Before we became apex species, we were prey: to lions, crocodiles, leopards, hyenas, snakes, and raptors. Evolution didn't merely teach us to run or hide—it sculpted our neural architecture. The amygdala, centre of fear processing, evolved under persistent predatory stress. Our startle reflex, our pattern detection, even our night vision are products of this pressure. *Fear of the predator isn't an emotion we learned—it's a neurochemical legacy we inherited.*

Cultural memory may encode this legacy in the form of monsters. Consider the recurring traits: fangs, fur, elongated limbs, glowing or reflective eyes, uncanny agility, nocturnal habits. These are not random. They are phenotypic approximations of real threats—selective exaggerations that ensure memorability. The point is not that there once was a literal werewolf, but that *the werewolf is a mnemonic vessel for the wolf*, exaggerated in form so as to secure attention, allegiance, and behavioural compliance in a community. Myths that encode risk are adaptive, and stories that sharpen vigilance have survival value. In this sense, the bogeyman is not fiction, but *heuristic technology*.

Dreams often reproduce these predator motifs with startling fidelity. Night terrors in children across cultures disproportionately feature being chased, attacked, or devoured by non-human figures. These are not dreams of falling or flying, but of being hunted. Psychiatrist Ernest Hartmann's research into nightmare content noted that predator-like chasers far outweighed any other threat archetype. Ethological studies suggest that animals, too, dream of predators—indicating that *the fear of being eaten is an architecture embedded not only in waking behaviour but also in the structure of consciousness itself.*

From a neurocognitive perspective, this suggests that predator archetypes aren't taught so much as triggered. In the right context—darkness, isolation, wilderness—they surface with remarkable ease, even in urbanised individuals. The unexplained rustle in the trees, the sudden silence of birds, the movement in the peripheral field—all reliably evoke heightened alertness, even in people who know, rationally, that no threat exists. This mismatch between cognition and reaction is not irrationality; it is *evolutionary preparedness*. Better to mistake wind for a panther than a panther for wind.

Legends, then, act as social simulations. They allow a group to run drills on scenarios too dangerous to test in real life. The tale of the shapeshifter in the woods, the flesh-eating spirit, the subterranean lurker—these serve to direct group norms, instil caution in children, and remind communities of the fragility of safety. And crucially, they preserve the *image* of threat in stylised, transmissible form. Language becomes a survival prosthetic, allowing individuals to inherit fear knowledge without direct exposure. *The child who learns the story of the aswang or the skinwalker has already internalised a partial model of threat-avoidance behaviour before encountering any real danger.*

There is also a darker dimension: the human predator. Some of the most enduring legends—vampires, ghouls, incubi, ogres—mirror not animal threats, but aberrant human behaviours: stalking, cannibalism, predation upon the vulnerable. These myths are not about external threats but internal dangers—members of the tribe gone rogue. Such stories perform a dual function: warning against specific behaviours, and marking those behaviours as ontologically monstrous. The man who preys upon women, the elder who harms children, the stranger who violates kinship norms—these are not merely immoral, but *bestial*, placed outside the moral community through mythic taxonomy. In this way, the predator archetype polices social boundaries as much as physical ones.

Yet the persistence of these forms suggests more than utility. They appear not just in morality tales or cautionary legends but in hallucination, psychosis, and the waking visions of trauma survivors. People who experience post-traumatic stress disorder often describe figures that feel distinctly non-human—eyes in the dark, shadow men, shape-shifting attackers. These are not symbolic, but *visceral returns of ancestral pattern recognition systems misfiring under duress*. The underdream—this subconscious substrate of predator imagery—surfaces most often not in myth, but in crisis. And when it does, it bypasses narrative entirely, erupting into pure, affective experience.

What makes this particularly compelling is how consistent the visual motifs remain, even across separated populations. Take the phenomenon of the "hat man"—a dark, humanoid figure often reported during sleep paralysis episodes. He appears with astonishing consistency across cultures that share no common mythology. The same is true for descriptions of skinwalkers, djinn, or night demons: all present as elongated, fast, often bipedal predators with ambiguous facial features and an aura of malevolence. This is unlikely to be cultural diffusion alone. It suggests a *shared visual lexicon of fear*, inscribed not through text or ritual but through millennia of survival under predation.

The mythic predator thus becomes not a lie we tell children, but a neurological artefact—a kind of species-wide trauma archive. It encodes not just memory, but instruction. It says: here is the shape of the thing that once hunted you. Do not forget. Do not dismiss. And though the real predator may be extinct, the form remains relevant—useful even—because the risk landscape has changed but not vanished. The predator has become metaphor: for power, corruption, addiction, abuse. The vampire is now a narcissist, the siren a manipulative algorithm, the ogre a warmonger in a suit. But the logic of predation—the strong consuming the weak, the cunning exploiting the naïve—remains unchanged. The predator form persists because it remains *adaptive to recognise.*

What does it mean, then, to live with these underdreams still pulsing beneath our digital overlays? It means we must take folklore seriously—not as literal belief, but as diagnostic residue. Myths can be read as collective MRIs: scans of our deep-time fears rendered in narrative. They reveal where trauma has pooled, where vigilance still anchors thought, where ancient categories still frame new threats. The underdream is not superstition; it is infrastructure.

To dismiss legendary monsters as mere fictions is to miss their function. They are not meant to be believed, but *understood.* The predator in the story is not the problem. It is the alert system, the rehearsal, the encoded message: remember what danger looks like. In a world where threats now wear human faces, corporate logos, or neural networks, the predator mythos teaches us how to *see what wants to remain hidden.* And in doing so, it reminds us that evolution did not end when we invented language. It simply moved indoors—into memory, metaphor, and myth.

Chapter 6
Gods as Technology — Myths of Power and Machines:

There is a provocative notion—circulated in the margins of alternative histories and lurking in the undercurrents of old epics—that myths may not be metaphors at all, but garbled recollections of technology. Those flying chariots, thunderbolts, speaking statues, and weapons that breathed fire might not be the fruits of symbolic imagination, but the faded echoes of real machines. To contemporary sensibilities, it may sound eccentric, even fanciful. Yet remove the insulating comfort of modern invention and a disquieting question emerges: how would people with no technology of their own recognise it? The answer, again and again across civilisations, is simple: they would call it divine.

Across the mythic canons of the world, there are stories of beings who descend from the skies armed with tools of uncanny precision and force. They shape mountains, command time, harness the elements as weapons, and vanish when their work is done. To the ancients, such beings were gods. To a modern ear, they can sound suspiciously like engineers, pilots, or something more advanced still. This is not a blunt claim that deities were extraterrestrials, but rather an invitation to see myths as archival accounts of encounters with powers so far beyond the local norm that they were indistinguishable from magic—powers which, in some cases, may have been technological in origin before being embroidered by centuries of storytelling.

Take the vimanas of ancient India. These airborne craft, found in Vedic and later Sanskrit sources, are described with a technical granularity that sits awkwardly with the idea of mere allegory. Accounts detail propulsion methods, aerial manoeuvres, landing techniques, even armaments. The Vaimanika Shastra—however disputed in provenance—offers a body of knowledge framed entirely around such machines. Whether these are descriptions of actual devices or elaborate containers for half-remembered knowledge, the specificity is striking. To imagine such vehicles wholesale would require a mechanical vocabulary and a grasp of aerodynamics well beyond what is usually credited to pre-industrial cultures.

The same motif recurs elsewhere: the god-borne craft of Gilgamesh, the interlocking wheels of Ezekiel's vision, the flying "shields" in Livy's Rome, the sky-dragons of Han Dynasty chronicles. If read not as religious fictions but as reports translated through the conceptual limits of their time, these accounts invite a radical reinterpretation. The beings they describe need not be alien in the literal sense; they could have been other humans, or even other intelligences, wielding tools entirely opaque to those who saw them. In that light, myth is not nature explained away—it is the residue of direct contact with overwhelming difference.

Such contact forces the mind into an adaptive manoeuvre. When faced with phenomena that rupture its conceptual frame—light without fire, flight without wings, motion without visible force—the human brain does not simply reject the input; it recasts it. It wraps the unassimilable in the cloth of divinity. This reflex preserves the experience without demanding an explanation that the culture cannot yet supply. Over generations, these preserved anomalies solidify into pantheons, epic cycles, miracle traditions. They are remembered, but their origin is buried.

And there is often, beneath these stories, a note of loss. Time and again, the gods are said to have brought gifts—fire, agriculture, metallurgy, governance—only to depart. The departure itself becomes a central refrain. From the Sumerian laments over Inanna's descent, to the Hopi tales of the Kachinas' retreat, to the Mesoamerican legend of Quetzalcoatl's sea-bound exit, the narrative is the same: those who came from the sky left again. Such repetition hints at something more than poetic closure. The "gods leaving" might be the mythic form of a post-contact rupture: the sudden absence of a guiding force, real or perceived, and the knowledge—perhaps infrastructure—that vanished with them.

From a psychological standpoint, the god-figure in such tales is a projection screen for awe and fear alike. Their weapons—the vajra of Indra, Zeus's thunderbolt, Shiva's trident—are not mere metaphors, but archetypal tools of dominion. They embody what some might call the technological sublime: the visceral shock of encountering forces that dwarf human scale. In this sense, the terror and reverence they inspire is not a reaction to "magic" but to systems so far beyond linear causality that they defy comprehension.

If ancient minds projected machines onto gods, modern minds may be reversing the equation—projecting godlike qualities onto machines. Artificial intelligence, for example, is steadily acquiring the attributes once reserved for deity: omniscience, unblinking memory, decision-making beyond human grasp. The "singularity" is imagined like a return of the gods, and fears of superintelligence echo the dread of an amoral divine will. The pattern repeats: technology begets gods; gods become the language for technology. Studying divine machinery in myth, then, becomes an act of studying ourselves—our impulse to sanctify what we do not fully understand.

Even the fashionable speculations of simulation theory fall into this old pattern. If our universe is a coded construct, then the coders are effectively gods, shaping reality's rules from outside. The fact that such ideas arise from the minds of software engineers and physicists is telling: theology and technology have converged again, and the oldest questions—Who made us? Why?—resurface in a digital key.

Thus, the ancient visions of godlike machines are not relics of a credulous past; they may be prototypes for our future encounters. They chart the human response to intelligences that act faster, think differently, and work with tools opaque to us. Once, those encounters birthed religions; now, they might give rise to ethical dilemmas in AI, contested space treaties, or existential debates over extraterrestrial contact. If myths are indeed encoded reports of such meetings, then reading them carefully may be less an exercise in nostalgia than preparation.

This line of thought will take us into multiple terrains: examining the "flying machines" and advanced weapons of ancient literature as potential distortions of real events; tracing the motif of divine withdrawal as the cultural scar of lost contact; weighing whether some gods were projections onto alien or artificial minds; and exploring how humans instinctively deify technological superiority.

The aim is not to declare one definitive truth, but to keep the stakes in sight. Should myths prove to be memories of technology, then history is not a smooth ascent but a jagged, interrupted experiment. If gods were once the names we gave to machines, then religion itself becomes a record of invention. And if we are now building tools with those same attributes, we must ask whether we are becoming the new gods—or merely walking, eyes open, into their old mistakes.

To take these questions seriously is not to drift into mysticism. It is to read myth as a coded archive: a pre-scientific intelligence report, recorded in verse because prose had not yet the strength. In decoding it, we inherit the awe of those who saw more than they could name. And within that awe lies, as ever, both a warning and a promise.

"Flying Chariots and Divine Weapons":

When ancient peoples looked to the sky and witnessed something unexplainable—light without fire, thunder without storm, motion without animal—they didn't respond with denial. They mythologised. But mythologising isn't fabricating. It is a cognitive strategy for preserving anomaly. In many cases, it appears to be a memory structure that encodes not what they imagined, but what they *couldn't categorise.* Across civilisations—from the Sanskrit epics of India to the battle hymns of the Norse skalds—we find references to flying vehicles, lightning-like weapons, self-directed arrows, and sound-based armaments. These are not scattered embellishments; they are repeated motifs. If we take the descriptive precision of some of these accounts seriously, they begin to read less like metaphor and more like misfiled technology.

In the *Mahabharata* and *Ramayana*, the Vedic texts of ancient India, the vimanas are flying chariots, described in such detail that some researchers have posited they might reflect an advanced, now-lost technological understanding. These vimanas are not portrayed as spiritual vehicles; they are outfitted with weapon systems, navigational controls, and materials that resist fire. One section describes a weapon called the *Brahmastra*—a projectile so powerful it causes wind to roar, birds to fall from the sky, and survivors to experience hair and nail loss. That constellation of effects—blast wave, avian disorientation, radiation sickness—*reads disturbingly like a post-nuclear scene*, though the cultural language of the time had no atomic vocabulary to explain it.

Sceptics often argue that these descriptions are merely allegories for natural forces or inner states. Yet this reading requires more contortion than it avoids. Why do we consistently find mythological accounts across geographically isolated cultures that describe sky-borne vessels emitting fire, sound, or light—often in combat scenarios, often involving "gods" who wield these tools selectively? When Gilgamesh seeks the secret of immortality in the Sumerian epic, he encounters a being with power over the sky. The Book of Ezekiel describes a vision involving wheels of fire and beings emerging from a radiant craft, complete with gyroscopic movement and multi-directional locomotion. The Popol Vuh of the K'iche' Maya describes gods of thunder and lightning who could split mountains and communicate across vast distances. None of these are identical—but they rhyme. And *that rhyme is worth interrogating.*

There is a critical distinction to make here. Suggesting that ancient peoples might have witnessed, interacted with, or inherited knowledge of high technology is not a dismissal of their cultural brilliance. On the contrary—it treats their observations with dignity. It assumes they reported extraordinary phenomena with the best linguistic and conceptual tools available to them. To wave away all such accounts as either symbolic or delusional is a form of chronological arrogance. It implies modern categories are the only valid ones, and ancient people were incapable of discernment. *But what if they were simply describing phenomena their epistemology had no lexicon for?*

This view invites a deeper analysis of mythological weapons. Take Zeus's thunderbolt. Often depicted as a lightning-shaped spear hurled from Olympus, it kills instantly, with blinding light and reverberating sound. Is it pure meteorological metaphor, or a crude rendering of a directed-energy weapon? The Norse god Thor wields Mjölnir, a hammer that returns to his hand after being thrown and causes thunder when it strikes. To the modern reader, this evokes a self-guiding missile system with immense kinetic force. In both cases, the weapon is not only powerful but *characteristically beyond the user's physical capacity to create*—it is often gifted by dwarves, gods, or cosmic forces. This implies it is not a product of human technology, but an import. And herein lies a conceptual tension: mythic weapons are both described in material terms and disavowed as human artefacts. That duality may hint at remembered technology from outside the immediate culture—either inherited, discovered, or witnessed.

It is tempting to categorise these motifs under Jungian archetypes—the sky god, the thunder weapon, the sacred vehicle. And archetypes are indeed relevant. But even archetypes have roots. They are not spontaneous inventions of the psyche. They are consolidations of recurrent experience, encoded as psychic constants. If flying chariots appear in Babylonian, Egyptian, Indian, Chinese, and Mesoamerican sources, that suggests not only symbolic convergence but *experiential commonality*. Human memory does not cling to irrelevance across millennia. What endures is what shocks, wounds, or empowers. So when we see cultural motifs persisting with fidelity—descriptions of beings in aerial platforms, of energy weapons devastating entire armies—it warrants more than literary interpretation. It demands technological reading.

This approach does not require belief in aliens. It only requires belief in *history's incompleteness*. There may have been civilisations, technologies, or intelligences now erased by time, tectonics, or systemic forgetting. Just as the Antikythera Mechanism redefined our understanding of Greek mechanical sophistication, or the Baghdad Battery challenged assumptions about Mesopotamian electricity, these mythic accounts may serve as *testaments to lost peaks of technical capacity*. If such tools existed—even temporarily—they would have appeared divine to those who had never seen metal shaped into flight.

Modern comparisons sharpen the picture. A 21st-century drone strike would be indistinguishable from divine wrath to an Iron Age observer. The sound signature, aerial trajectory, and destructive power of even a small missile evoke supernatural association. Similarly, modern weapon systems—laser-guided bombs, kinetic railguns, autonomous aerial vehicles—are closer in behaviour to mythic artefacts than to traditional human weapons. They are wielded remotely, often unseen, and strike with a precision that seems uncanny. In this light, ancient descriptions of divine weapons may not be exaggerations—they may be analogues.

The argument that technology shapes mythology is not new, but it has been sidelined in academic circles due to its association with pseudoscience. Yet this guilt-by-association has stifled rigorous exploration. It is possible to treat ancient texts with both critical distance and interpretive generosity. To read the *Mahabharata*, *Enuma Elish*, or the *Kojiki* as both mythic and empirical documents is not to lose scholarly credibility—it is to expand its boundary conditions. *After all, archaeology has repeatedly vindicated what philologists once dismissed as fantasy.*

The practical implications are significant. If some mythological artefacts represent misunderstood technology, it reframes the study of myth as a repository of technological anthropology. It also forces contemporary cultures to reassess the linear model of technological development. Perhaps the rise of machines is not a straight line, but a cycle. Civilisations may gain, lose, and mythologise power repeatedly, depending on environmental, political, or cosmic disruptions. The story of flying chariots and divine weapons, then, becomes a story not of fantasy, but of civilisational memory—partial, poetic, and encrypted.

This chapter takes this possibility seriously. It does not claim all myth is data. Nor does it reduce rich traditions to machine catalogues. But it *does* insist that the precision of some ancient descriptions—their technical specificity, their behavioural accuracy—deserves more than symbolic interpretation. It deserves multidisciplinary analysis: linguistic, anthropological, technological, and philosophical. At its best, this inquiry does not demystify myth. It honours it. It allows us to see that the ancients may have known more, seen more, or remembered more than we have given them credit for.

To return to the original question—what does high technology look like to people without any?—the answer remains: *it looks like a god.* But that does not mean it was one. It may simply mean it was a visitor, a relic, a survivor, or a machine. And it was remembered the only way available to pre-modern minds: in story. In fire and light. In thunder that carried names. In chariots that flew.

"The Vimanas of India and Their Modern Echoes":

In the annals of ancient Indian literature, few subjects incite as much fascination—and as much controversy—as the vimanas. These are not dreamlike metaphors or poetic flourishes. They are described, often with exacting specificity, as aerial vehicles. They appear in epics like the *Mahabharata* and the *Ramayana*, in Puranic lore, and in later Sanskrit treatises such as the *Vaimanika Shastra.* Depending on which text one reads, vimanas are airborne chariots drawn by divine beings, self-propelled craft with mercury-based propulsion systems, or geometrically precise flying machines capable of vertical take-off and long-range flight. The deeper one ventures into these descriptions, the less comfortably they sit within the tidy bounds of literary metaphor. Instead, they begin to resemble something altogether more unsettling: technological artefacts misplaced in time.

The temptation is to interpret these references allegorically—as symbols of spiritual ascent, cosmic journeying, or metaphorical warfare. But the degree of mechanical language in some texts resists easy poetic reduction. In the *Ramayana*, the Pushpaka Vimana is described as an aerial chariot "that resembles the sun and moves at the speed of the wind," capable of enlarging or shrinking at will and flying at great altitudes. In isolation, such a description might pass as artistic embellishment. But the corpus of Vedic and post-Vedic references to flying machines is vast, internally consistent, and culturally significant. They do not appear once or twice in mythological sidebars. They appear repeatedly and structurally, embedded in the narratives of gods, sages, and kings.

The *Vaimanika Shastra*, a Sanskrit treatise of uncertain provenance, purports to be a manual for constructing vimanas, offering detailed schematics and engineering concepts. It describes propulsion by anti-gravity mechanisms, engines powered by liquid mercury, and control systems for vertical and lateral movement. While mainstream historians rightly question the text's historical dating—most evidence suggests it was compiled or transcribed in the early 20th century—its detailed technical vocabulary cannot be easily dismissed as mere pastiche. At the very least, it signals a persistent and sophisticated tradition of imagining flight in pre-modern India. And if the *Vaimanika Shastra* is a modern fabrication, it is one informed by a tradition with deeper roots. It did not appear from a vacuum.

To understand vimanas properly, one must sidestep both the credulity of the pseudoscientific fringe and the reflexive scepticism of academic orthodoxy. Both extremes obscure the more difficult, intellectually demanding task: to ask *why* advanced aerial vehicles were such a consistent theme in ancient Indian texts, and what that reveals about cultural memory, epistemological frameworks, and technological possibility. If vimanas are symbolic, *what exactly are they symbols of?* And why do they so often appear in contexts where power, mobility, and sovereignty are at stake?

There is a case to be made that the vimana is a vessel of memory—an encoded relic of something once seen or half-remembered, misinterpreted or translated across centuries of oral tradition. If a technologically advanced object were observed by a pre-industrial society—especially one deeply attuned to cosmology and the sacred—it would likely be absorbed into myth rather than chronicled as engineering. This is not a flaw of the culture; it is a cognitive inevitability. Societies describe anomalies in the terms available to them. To dismiss such accounts simply because they lack modern technical framing is to apply a historically naive filter to what may be legitimate ethnographic data.

The scholarly discomfort around vimanas often stems from a larger anxiety: the threat they pose to the linear model of technological history. If the ancients possessed—or even observed—flight-capable technology, the canonical narrative of human progress as a steady climb from ignorance to enlightenment begins to wobble. But the archaeological record has already shown that civilisational peaks are not linear. Sophistication rises and collapses in cycles. Consider the Antikythera Mechanism: a device from 2nd-century BCE Greece that functioned as an astronomical computer with astonishing accuracy. It predated anything similar by over a thousand years and was entirely out of place in its technological context. Its discovery forced a revision of assumptions about ancient Greek engineering.

Might the vimanas, if taken seriously, prompt a similar reconsideration of ancient Indian technological philosophy?

The parallel with modern aerospace technology is striking. Several 20th-century and early 21st-century thinkers—including aerospace engineers, physicists, and historians of science—have noted uncanny resemblances between vimana descriptions and principles of modern flight. Concepts like gyroscopic balance, centripetal propulsion, and anti-gravity—while wrapped in metaphysical language—appear in early Sanskrit materials. The challenge is interpretive. Are these descriptions literal, metaphorical, or distorted memory? And more importantly, *can these categories even be meaningfully separated* in a culture where the sacred and the empirical were never disentangled?

The use of mercury, often cited in vimana propulsion narratives, is particularly curious. Mercury's properties—its density, mobility, and responsiveness to electromagnetic fields—have long fascinated alchemists and engineers alike. In modern aerospace contexts, mercury has been explored (though ultimately discarded) as a component in certain experimental propulsion systems. The fact that mercury is named so frequently in ancient vimana accounts may suggest either coincidental symbolism—or ancestral knowledge of its potential relevance to flight mechanics. In either case, the consistency is hard to ignore.

Critics argue that modern readings of vimanas are simply retro-projections—that we impose technological interpretations onto texts that never intended them. But this critique presupposes that symbolic and technological readings are mutually exclusive. They are not. The human mind encodes technology through narrative just as it encodes myth through metaphor. The ancient Indian mind was no less capable of layered description than ours. Indeed, much of Indian cosmology operates on precisely this dual register: literal and symbolic, material and spiritual. To insist on a binary is to misread the culture.

What makes the vimana question especially provocative is its recurrence in comparative mythology. Other cultures also contain persistent references to flying craft, sky-borne battles, and artefacts that defy gravitational logic. The Chinese speak of dragons that could ascend to the heavens and travel vast distances with riders. The ancient Egyptians revered Ra's solar barque, which traversed the sky and underworld with luminous speed. The Norse gods crossed realms via Bifröst, a luminous bridge or beam of some sort. When taken collectively, these motifs suggest more than isolated inventions. They may be memory traces—glimpses of a civilisational layer that lies beneath our own recorded history, long erased but partially preserved in story.

There is also a political dimension to the vimana discourse. In post-colonial India, the vimana has often been invoked as a symbol of indigenous scientific pride. This has created tension between legitimate exploration of ancient technological concepts and nationalist myth-making. While it is crucial to disentangle scholarship from ideology, one must also recognise that dismissing these traditions outright can itself become an ideological act—a form of academic gatekeeping that enforces Eurocentric assumptions about the chronology of innovation.

Modern interest in vimanas has led to attempts at reverse-engineering their principles using digital modelling, fluid dynamics, and electromagnetic simulation. While none have yet produced a working prototype, the thought experiment alone is revealing. It shows that ancient descriptions, when stripped of their religious overlay, often suggest plausible aeronautical principles. That fact alone justifies serious multidisciplinary inquiry—not to prove ancient aviation, but to expand the boundaries of what we allow ourselves to consider historically credible.

This subchapter does not argue that vimanas were real in the literalist, physical sense. It argues that their recurrence, specificity, and cultural centrality point to something real in the cognitive sense—something seen, imagined, remembered, or misremembered. That "something" may have been a meteorological phenomenon, an advanced artefact, a lost technology, or a cross-cultural hallucination. But dismissing it out of hand because it does not fit into modern models of plausibility is a failure not of evidence, but of imagination.

To engage seriously with the vimana tradition is not to abandon science; it is to practise a broader, more inclusive archaeology of knowledge—one that respects the intelligence of ancient observers and treats myth as a source of encrypted data. Whether vimanas were machines, metaphors, or memories, they compel us to reconsider what ancient civilisations may have known, and how they chose to remember it. That, in itself, is a form of technological insight—one no less powerful than the machines we now send into our own skies.

"Alienation and the God-Complex":

When myth places a divine figure in the sky, descending in fire or light to deliver wisdom, punishment, or transformation, it reveals less about the heavens and more about the human psyche. The gods of antiquity are often framed as literal beings—charioteers of clouds, wielders of thunder, masters of death and fertility—but their narrative function stretches far beyond cosmology. They are carriers of psychological weight. The overwhelming consistency across global myths of advanced, often alien-like deities—beings imbued with powers ungraspable to their human observers—demands not only comparative mythography, but psychoanalytical scrutiny. These entities may not be visitors from another star system. They may be projections of the most disquieting parts of ourselves: our awe of power, our fear of abandonment, our craving for meaning, our compulsion to explain suffering.

The gods of myth are not just creators; they are interrupters. They arrive uninvited. They wield disruptive power—imposing laws, destroying cities, demanding sacrifice. They manifest suddenly in moments of civilisational rupture: the flood, the tower, the fire from the sky. Consider how this structure mirrors the way trauma interrupts the continuity of personal narrative. Many of these divine beings resemble what psychoanalyst Carl Jung called *archetypes of the Self*, emergent when the ego encounters its limits. The divine, in this context, is a psychical force: not a supernatural being, but a psychic overload—a phenomenon too large to integrate, too powerful to ignore.

There is a crucial pattern in these narratives: the divine figure appears not merely as 'other' but as *radically other*—exceeding human proportion in power, scale, and detachment. The alienness is essential. It marks a limit condition, a horizon of understanding. This radical otherness is what philosopher Emmanuel Levinas identified as the encounter with the "face of the Other," which ruptures self-certainty. Myths of advanced beings are encounters with the ultimate "other" in narrative form—beings who see without being seen, who know without being known.

In this sense, the "god-complex" is not merely a feature of the gods. It is a psychological condition in their human observers. To confront a being that transcends moral frameworks, disregards reciprocity, or exhibits disinterest in human suffering provokes a crisis of agency. Ancient humans responded by building temples and rituals—systems to contain and domesticate the incomprehensible. But what if these systems merely encoded a deeper existential dissonance: that we once *felt* in the presence of something beyond ourselves, but could neither classify nor control it?

The overlap between deities and so-called "ancient aliens" is not accidental. It reflects a modern reinterpretation of the same cognitive structure: we continue to externalise what we cannot fathom, to cast the technologically advanced or morally incomprehensible as "from elsewhere." The alien gods of Mesopotamia, the shining beings of Mesoamerican myth, the Vimanas of Sanskrit epic—each shares this essential trait: *they do not belong*, yet they intervene. They intrude. They alter the trajectory of the human world. This sense of intrusion is what grants them narrative and symbolic power. It also reflects a deep psychological response to the unknown: to render mystery in the shape of authority.

One of the most enduring motifs in such stories is the god who departs. After a golden age, or a cataclysm, or a betrayal, the deity leaves—ascending to the heavens, disappearing into the sea, promising to return at the end of days. This absence is haunting. It leaves a civilisational scar: the sense that the best is behind us, that knowledge has been lost, that we are abandoned children of some great ancestral power. The god's departure marks not only the end of contact, but the beginning of myth. That rupture—of contact followed by absence—is central to the structure of both religion and alien lore. It is the origin story of estrangement.

This dynamic also maps onto psychological structures of repression. The god may not merely represent power. It may also symbolise a *disowned part of ourselves*. Jung referred to the shadow—the unacknowledged, suppressed aspects of the psyche that we externalise as enemies or deities. The god becomes a canvas for the unbearable: violent justice, ecstatic wisdom, the power to destroy and create without moral ambiguity. These are traits human societies often cannot integrate, so we exile them into the realm of the divine.

Indeed, many of these gods exhibit what we would now diagnose as pathologies: they are wrathful, jealous, amoral, inconsistent. They are the tyrants of the unconscious. But rather than dismiss them as outdated constructs, we might consider what functions they performed for early civilisations. In a world lacking psychological language, the best mechanism for confronting complexity was narrative embodiment. To make a god of one's terror is to give it shape. To make a god of one's longing is to give it voice.

Contemporary media continues to recycle these tropes. From films like *Prometheus* to ancient astronaut documentaries, the notion of technologically advanced beings with godlike abilities and distant motives remains culturally magnetic. But the emphasis has shifted—from reverence to suspicion. Our modern god-complex often inverts itself: we are no longer awed by deities; we are obsessed with the *deconstruction* of godhood. We no longer want gods to intervene; we want to prove they never existed. Yet the need for them—psychologically, culturally—persists. This tension defines much of contemporary mythmaking: we reject the divine while mimicking it through technology, surveillance, and artificial intelligence.

This raises a vital philosophical question: when we project "advanced beings" into the mythic past, are we describing visitors from elsewhere—or are we anticipating our own potential future? The god-complex may be a mirror, not a memory. As our technologies approach powers once attributed to deities—gene editing, weaponised weather, synthetic consciousness—we begin to retroactively mythologise our ancestors as having once possessed similar abilities. This feedback loop between memory and projection complicates the historical reading of myth. It suggests that our ancestors may have embedded anticipatory visions of our present within the structure of their sacred stories.

Consider also how these myths function socially. The gods are rarely solitary. They are part of pantheons—hierarchies of power that resemble human political orders. This suggests not just projection of psychological structures, but *political* ones. Myth, in this context, becomes a vehicle for discussing sovereignty, legitimacy, and obedience. When a king claims divine descent, he is importing the god-complex into governance. When a priest claims access to divine will, he becomes the interface between human and transcendent power. This divinisation of hierarchy continues in modern secular forms—states, ideologies, and technologies that assert control through narratives of superior knowledge.

The most unsettling implication, however, is epistemological. Myths of advanced beings may not be distortions of memory or predictions of progress, but insights into a different kind of knowledge altogether—*liminal knowledge*, existing on the edges of articulation. These beings do not operate within our categories of true and false, fact and fiction. They belong to what philosopher Paul Ricoeur called the "second naïveté": not the literal belief in myth, but a conscious participation in its symbolic world. In this light, the god-complex is not a bug of human cognition. It is a feature—an adaptive mechanism for metabolising the incomprehensible.

To reduce the gods to primitive science-fiction is to flatten their symbolic power. But to treat them purely as metaphors is equally myopic. They are both: metaphors built from memory, and memories shaped by metaphor. Their alienness speaks to our alienation—not only from the cosmos, but from aspects of ourselves we fear to recognise. In creating gods, we perform an act of self-displacement: *we place our greatest potentials—and our greatest fears—into beings we can worship, obey, or destroy without admitting they come from within.*

This subchapter does not aim to resolve the question of whether ancient gods were astronauts, archetypes, or allegories. Rather, it explores how each of these possibilities points to a deeper truth: that the divine, in myth, is rarely about the gods. It is about us—our limits, our longings, and the stories we tell when faced with the enormity of what we do not yet understand. To confront the god-complex is to confront our own complexity, and to recognise in myth not superstition, but the architecture of human thought stretched to its highest imaginative reach.

"When the Gods Left":

There is a recurring motif in ancient mythology that quietly unsettles: the gods not only arrive—they also leave. In culture after culture, from Mesoamerica to Mesopotamia, stories endure of divine beings who once walked among humans, imparted knowledge, altered the natural order, and then vanished. Sometimes they ascend to the sky, sometimes they disappear into the sea, sometimes they simply turn away. The pattern is so persistent—and so underexamined—that it raises a profound question: *were these narratives simply theological metaphors for natural cycles and lost hope, or are they refracted memories of actual contact with advanced, now-absent civilisations?*

In Aztec cosmology, Quetzalcoatl, the feathered serpent deity associated with learning, science, and wind, is said to have departed across the ocean, vowing to return. His departure was not framed as death or transformation, but as an *exit.* The very expectation of his return shaped the psyche of later rulers and even intersected fatally with the arrival of Hernán Cortés. In Sumerian myth, the Anunnaki gods once ruled directly over human city-states before withdrawing to the heavens. In Polynesian tradition, god-ancestors such as Kane or Tane retreat after bestowing order. In Hindu cosmology, the *Devas* interact directly with humanity in earlier *yugas*, or world ages, before withdrawing in later, more degraded times. The message is consistent: the presence of gods was not always distant—it was proximate, active, even political. And then, it stopped.

These myths are too widespread—and too specific in their structure—to be dismissed as coincidence. More importantly, they are not uniformly redemptive or hopeful. The departure of the gods is often marked by trauma: flood, fire, civil collapse, cosmic war. The motif is not simply about celestial withdrawal; it is about *catastrophic severance.* What follows is a loss of knowledge, a fall into forgetfulness, and a nostalgic yearning for recontact. This trope strongly suggests that the stories encode more than theological doctrine. They carry an emotional weight akin to the cultural memory of disaster.

The structure of these stories matches that of what cognitive anthropologists refer to as *post-contact myths*: legends that arise after a society encounters a foreign power, experiences a moment of overwhelming disruption, and then mythologises the event through a lens of sacred narrative. These myths serve to explain what has been lost and provide a scaffolding for surviving identity. Could it be, then, that the so-called departure of the gods encodes a much older form of post-contact memory—one reaching back to forgotten encounters with non-human intelligence, or lost human civilisations possessing radically advanced technology?

This hypothesis requires neither extraterrestrials nor fantasy. It simply invites us to treat ancient testimony with methodological seriousness. Take the Dogon people of Mali. Anthropologists Marcel Griaule and Germaine Dieterlen recorded, in the mid-20th century, that the Dogon preserved an oral tradition about a star system—Sirius A and Sirius B—describing properties of its companion star that were only verifiable with modern telescopes. While some critics have argued that such knowledge could have been introduced earlier by missionaries or colonial administrators, the more critical question remains: *why do so many cultures encode anomalous scientific data in the mouths of gods who have left?* If these stories were merely inventions, why embed such esoteric astronomical detail in divine departure narratives?

Consider also the Popol Vuh of the K'iche' Maya. In it, gods create multiple versions of humanity before retreating into the sky, leaving behind only fragmented knowledge. The final humans, the ones we know today, are made of maize, the sacred crop, but they lack the full vision and understanding of earlier iterations. They are, in a sense, the degraded inheritors of a more complete past. The gods do not simply vanish; they withdraw *in protest*, leaving a world that no longer meets their standard of harmony. In this frame, divine absence is not natural—it is punitive.

Modern psychological theory offers an additional layer of insight. When children experience abandonment, they often develop coping mechanisms to interpret or justify the absence of a caregiver. These can take mythic form—stories of being left behind for one's own good, or of a parent promising to return. Cultures, too, engage in mythopoeic compensation. The idea that a superior being once guided us, but left us with a promise of return, offers comfort amid disorientation. It recasts a traumatic break as a divine plan. If the gods left, then we are not abandoned—we are awaiting reunion.

Yet the symbolism may run deeper still. The withdrawal of the gods often coincides with the rise of human kingship. In this transition, power moves from the divine to the political. The early rulers of Sumer, Egypt, and Mesoamerica were not simply sovereigns—they were mediators of the absent divine. The Pharaoh was the *Ka* of Ra; the Inca was the child of Inti; the Sumerian king was appointed by Anu. Such roles make little sense unless the gods had once been directly present. Rulers step into the vacuum left behind. The divine exit, in this frame, is not just myth—it is a legitimising structure for human authority. But the tension persists: the memory of the gods as actual presences haunts these regimes, producing anxiety about legitimacy, succession, and the loss of the sacred.

There are archaeological implications to consider. Sites such as Göbekli Tepe, Teotihuacan, and the Indus Valley display levels of sophistication in urban planning, astronomical alignment, and symbolic abstraction that seem anomalous for their timeframes—especially when considered in relation to the technological discontinuity that follows them. Often, these civilisational peaks are followed not by incremental advancement, but by collapse, abandonment, or mysterious interruption. The absence of a clear transmission of knowledge from these sites to later cultures suggests a discontinuity. The idea that certain advanced practices were brought by beings—divine, semi-divine, or simply highly advanced—and then lost, fits both the mythic and material record.

Critics rightly warn against the temptation to fill every cultural gap with speculative theory. However, the discipline of cultural memory studies teaches us that myths are not fabrications; they are *codifications*. They preserve affective truths, not literal timelines. When multiple civilisations independently narrate a god's retreat, followed by disarray, it reflects a consistent experiential pattern: something vital *was once present*, and is no longer. This does not demand belief in gods. It demands belief in the human capacity to remember trauma creatively.

There is also a metaphysical reading. The divine departure may not describe a historical event, but a shift in consciousness. In archaic societies, the boundaries between self, cosmos, and spirit were porous. Myth may describe a world in which the divine could be directly felt—an animistic, intersubjective cosmos. As urbanisation and abstraction increased, those experiences became rarer, harder to interpret, and thus mythologised as something lost. The gods did not leave; we lost the means to perceive them. In this reading, divine withdrawal is not spatial—it is perceptual.

A final tension deserves mention. The gods' departure often carries a caveat: *they will return.* Whether in apocalyptic visions, messianic prophecies, or golden age restorations, myth cycles promise the return of the sacred. This eschatological framing is not merely religious. It speaks to a psychological structure that resists accepting loss as final. The notion that the divine will return restores teleology to a world that might otherwise feel accidental. Yet embedded in this hope is also a warning: when the gods return, they do not always come in peace. Their reappearance often heralds judgment, reckoning, and the violent restoration of cosmic order.

So what do these stories truly offer us? They suggest that myth remembers more than it reveals. They encourage us to reconsider the line between history and imagination, between sacred loss and structural change. Whether one sees the departing gods as metaphors for ecological collapse, fragments of ancestral trauma, or blurred memories of prehistoric encounters, the commonality across civilisations is intellectually inescapable. Something—or someone—left. And humanity has never stopped trying to explain why.

Perhaps, then, the most productive reading is not to ask whether the gods *were* here, but to ask what these stories tell us about what happens when a civilisation outgrows its own myths, yet still feels the absence of the truths they once encoded. The myths of divine departure endure because they express the one truth no culture can escape: *we are always living in the shadow of what we have lost.* Whether gods, knowledge, or innocence, something is missing. And it is the mark of intelligence—not superstition—to ask what that something was.

Chapter 7
Dreams, Prophets, and the Neuroscience of Revelation

Through the ages, across empires and fading kingdoms, the dream has been a contested territory—exalted as a message from the gods, dreaded as the symptom of a troubled mind, and, in recent centuries, scrutinised as a function of neurobiology. Yet even with the bright lamps of neuroscience casting their light, the dream retains a haze of sacred ambiguity. In sleep, we wander through landscapes we've never walked, meet beings we never consciously imagined, and witness events we could never have devised. Some dreams alter the course of a life; others give rise to religions. This chapter begins not with rigid doctrine nor speculative fancy, but with a simpler, more unsettling premise: dreams—and the visionary states that accompany them—may reveal the brain's native grammar for encountering the unknown, whether that unknown lies deep within or far beyond.

The stories themselves are well-worn, yet their implications often go unexamined. Ezekiel envisions a wheel nested within another, glittering like beryl, guided by creatures whose four faces gaze in every direction at once. Joan of Arc, at thirteen, hears the voices of saints, their commands ringing in her mind with such certainty that neither prison nor inquisitor could shake her resolve. We tend to frame such accounts in polar terms: miracles or madness, revelation or delusion. But such neat categories risk missing the more curious question: what is the brain doing when it fabricates a world more vivid than waking reality, richer in meaning than reason can contain, and more enduring than memory itself?

Among neuroscientists, there is growing acceptance that visionary states—whether occurring in epileptic seizures, psychotic breaks, or REM sleep—are not simply misfires of perception. They may represent evolved modes of consciousness, serving functions such as pattern recognition, emotional processing, or even psychological renewal. Temporal lobe epilepsy, for example, has been linked with heightened religiosity, complex visual imagery, and phantom voices.

This invites a provocative possibility: perhaps some historical prophets owed their visions less to divine visitation than to distinctive neuroelectric wiring. If so, to dismiss such experiences as mere pathology would be to overlook their potential function. What we call revelation may be less a breach from outside, more a reorganisation of the mind's deepest machinery.

Enter the pineal gland—an organ that has attracted both anatomical inquiry and mystical speculation for centuries. Descartes named it the "seat of the soul," imagining it as the bridge between thought and flesh. Today, physiologists know it chiefly as the regulator of melatonin and circadian rhythms. Yet symbolically it stands elsewhere: the "third eye" of esoteric lore, the organ of inner sight. In the dream state, when the gates to sensory input close, the pineal behaves like a faint, pulsing beacon—responsive to cycles of light even in total darkness. Though science has yet to pin down its precise role in imaginative states, its position and hormonal output make it a plausible player in the neurochemical theatre of dreaming. Here, biology and myth do not contradict each other; they speak in chorus.

Jungian psychology offers another perspective. Jung read recurring dream symbols—deluges, caverns, shapeshifting beasts—not as echoes of the day's detritus, but as signals from the collective unconscious, a storehouse of inherited imagery. These archetypes are not invented anew; they are passed down, the psychic DNA of our ancestors. Seen in this light, a prophetic dream is not a crystal-ball forecast, but a transmission from evolutionary memory: a mind rehearsing its responses to dangers and thresholds encountered over millennia. Take the dream of being hunted: from a survival standpoint, the imagery makes perfect sense. Those whose minds encoded danger into symbol were more likely to survive—and to pass down that pattern. The wild predator morphed into a demon; the untamed forest into the unconscious; the dream into a sacred text. We still read it, whether we realise it or not.

But there is yet another thread to follow: the prophet not as a chosen vessel, nor a traumatised visionary, but as a person whose cognition tilts toward patterns others overlook. The "oracle brain" hypothesis proposes that some individuals possess unusually interconnected neural networks, enabling rapid, cross-domain associations. The result might be acute intuition, a heightened facility with metaphor, or a capacity for experiences that hover between senses and categories. In earlier times, such traits might yield oracles, poets, or holy madmen; in modern terms, we might speak of cognitive divergence—a brain that doesn't merely predict reality, but bends its own models to see connections others miss. In this view, prophecy is not an aberration but an evolutionary trial run.

This chapter weaves these strands together—not to solve the puzzle of revelation, but to honour its complexity. We will consider how surges of neural activity can sharpen vision, how a small gland might become a lighthouse for myth, how dreams carry ancestral memory, and how certain cognitive architectures could give rise to a prophetic class. The aim is not to strip the sacred down to synapses, but to suggest that brain and spirit may be co-authors of the same text.

Such inquiry inevitably leads to the question: what exactly is a revelation? Is it the arrival of a message from beyond, or the surfacing of something long latent? Is it a fault-line in the brain's predictive model, or a rare permeability to realities usually kept at bay? And is its content—moral codes, warnings of disaster, calls to devotion—mere cultural residue, or the distilled logic of survival in human communities? Perhaps prophets were less anomalies than instruments of cultural equilibrium, their visions forged by neurological difference and social demand.

Yet the matter of authority intrudes. Why does one person's dream vanish into oblivion while another's becomes scripture? Why is one vision dismissed as madness and another embraced as divine law? Revelation is never purely psychological; it is political. Who claims the right to speak for the gods? Who decrees that a midnight voice was holy rather than disordered? Often the difference lies not in the experience itself, but in the surrounding theatre—timing, influence, charisma, and cultural readiness. Revelation, once again, is filtered through the mesh of power.

Still, one feature of the visionary state appears stubbornly consistent: the compulsion to speak. Whether in the deserts of antiquity or the city streets of the present, the prophet emerges with a message that feels not merely important but imperative—something that must be told, whatever the cost. This inner command raises one of the most fascinating questions of all: what does it mean for the brain to generate a voice it experiences as alien to itself? What evolutionary advantage might such a phenomenon serve?

The pages that follow will not pretend to settle these questions. Instead, they offer a framework for viewing dreams, visions, and divine voices as maps—rendered in narrative form—of the mind grappling with what resists comprehension. Whether sparked by illness, sleep, ritual, or silence, such states can yield insights that civilisations have long treated as gifts from beyond. Perhaps they are. Or perhaps they rise from the strata beneath consciousness, from ancestral memory and evolutionary reflex.

The prophet, then, is not a relic from a superstitious age, but a living case study in how humans metabolise chaos, make stories of the ineffable, and find order in the flood of experience. To study them is not to dissolve their mystery, but to face the possibility that the capacity for revelation is not an accident of biology—it is one of its most astonishing designs.

Visions and Epileptic Trance States Case studies:

The human brain, at its most stable, is a pattern-making engine. Yet when its electrical rhythms are disrupted—whether through trauma, neurological irregularity, or altered states—it doesn't simply fail. Instead, it reconfigures. Visionary experiences, once viewed as either sacred or symptomatic, now sit at the intersection of neuroscience, cultural memory, and psychological interpretation. Among the most provocative of these experiences are those associated with epileptic trance states. These are not hallucinations in the popular sense, nor delusions in the clinical one, but vivid perceptual realities with consistent internal structure—narratively coherent, symbolically dense, and often transformative for those who undergo them. What should we make of them?

The biblical prophet Ezekiel, writing during the Babylonian exile in the 6th century BCE, describes a vision that has puzzled, inspired, and divided interpreters for millennia. "I looked, and I saw a whirlwind coming from the north, a great cloud with fire flashing forth continually and a brightness around it," he writes. What follows is not just religious theatre, but a detailed description of spinning wheels intersecting in space, beings with multiple faces, and an overwhelming sense of divine presence. While theologians and artists have long fixated on the metaphysics of this account, neurologists have noted something else: Ezekiel's vision bears striking resemblance to accounts given by patients with temporal lobe epilepsy (TLE). Intense light, rotating geometric patterns, sensory overload, auditory messages, and a profound conviction of truth—these are classic features of a TLE-induced aura or trance state.

Temporal lobe epilepsy does not always result in convulsions. In many cases, it produces *experiential seizures*: moments of hyperlucid insight, déjà vu, auditory phenomena, or hallucinatory states that feel charged with cosmic or moral meaning. Psychiatrist Norman Geschwind famously described a syndrome associated with TLE, characterised by hypergraphia, deepened religious feelings, and an intense drive to seek metaphysical understanding. What's notable is not that Ezekiel may have had epilepsy, but that *his brain's atypical functioning allowed him to generate images that became the cornerstone of a prophetic tradition.* The medical fact does not erase the spiritual resonance—it *grounds* it, revealing the neurocognitive scaffolding upon which myth can be constructed.

Joan of Arc presents a different but equally instructive case. Beginning at age thirteen, she reported hearing voices— initially soft, then increasingly urgent—identified as those of Saint Michael, Saint Catherine, and Saint Margaret. These voices gave her detailed instructions, moral certainties, and eventually military strategies. Joan never doubted their source, even under threat of execution. Her trial transcripts reveal not only her theological fluency, but also her neurological singularity. Today, her experiences might be described under the umbrella of auditory verbal hallucinations (AVH), a phenomenon well documented in both clinical and non-clinical populations. Yet what distinguished Joan was not just the content of her visions, but her capacity to act upon them with courage, coherence, and moral clarity.

Here, a critical distinction must be made. The presence of hallucinations or visionary experiences does not equate to psychosis. Many individuals report auditory or visual experiences outside the norm without displaying other markers of mental illness. In some anthropological contexts, these experiences are valorised rather than pathologised. Shamanic cultures, for instance, frequently interpret trance states as tools for healing, diagnosis, or cosmological navigation. Joan's visions—and the political consequences they carried—can be seen as evidence not of illness, but of *exceptional cognitive integration*: her ability to translate internally generated content into externally coherent action.

This raises a broader question: what is the difference between hallucination and revelation? Is it merely the presence of consensus validation? Or is there a neurological threshold beyond which visionary content becomes pathological? From a neuroscientific standpoint, visionary states involve complex interactions between the temporal lobes (processing emotion and memory), the parietal lobes (spatial orientation), and the limbic system (emotional salience). In such states, internal imagery may be rendered with such vividness and certainty that it is experienced as externally real. The resulting vision is not "imagined" in the colloquial sense—it is *perceived*, with all the attendant force of waking reality.

But perception is not infallible. Nor is it random. Visionary content tends to be patterned, archetypal, and laden with culturally resonant symbols. This suggests that while the *trigger* for the experience may be idiosyncratic, the *form* it takes is often historically and psychologically consistent. A wheel within a wheel. A voice calling from the sky. A visitation by a radiant figure. These motifs recur not only across individuals, but across civilisations. They suggest a cognitive template for religious awe, woven into the architecture of the brain.

This leads to one of the most intriguing and controversial hypotheses in the study of mysticism: that the human brain has evolved *modules for religious experience*. Evolutionary psychologists like Pascal Boyer and Justin Barrett have argued that belief in agency, intentionality, and moral causality may be hardwired into the mind. Others, like Andrew Newberg, have used neuroimaging to track how the brain responds during prayer, meditation, and peak spiritual states. His studies reveal a decrease in activity in the parietal lobe (associated with self-location in space), alongside increased activity in the frontal lobes (associated with focus and meaning-making). The subjective result is a sense of merging with a greater whole, losing the boundaries of the self. It is precisely this state—commonly reported by prophets and mystics—that convinces the experiencer they have contacted something *beyond* themselves.

And yet, it is important not to reduce all visionary experience to brain chemistry. To do so would be to misunderstand the *function* of such states in human life. Whether triggered by epilepsy, meditation, fasting, or trauma, visions have long played a role in moral decision-making, communal cohesion, and existential orientation. Their content may be biologically mediated, but their *meaning* emerges from narrative integration. That is, a vision becomes revelation when it is interpreted, recorded, and woven into a cultural framework that affirms its significance. Without this interpretive scaffolding, the experience remains private and possibly fragmented. With it, the experience becomes *transpersonal*—capable of shaping law, inspiring art, or launching revolutions.

It is no coincidence that both Ezekiel and Joan operated within cultural systems that provided interpretive frameworks for their visions. In ancient Israel, the role of the prophet was established, albeit contested. In medieval France, saintly voices were legible within Christian theology. Thus, the visionary state may be best understood as a kind of *psychological infrastructure*—a set of tools that, under specific conditions, generate symbolic data, which is then interpreted through the lens of culture and need. This model accounts for both the neurocognitive and the mythopoetic dimensions of prophecy.

One must also acknowledge the political utility of visions. Throughout history, visionary claims have been used to legitimate power, resist authority, or authorise violence. Prophetic speech can threaten institutional hierarchies, precisely because it invokes a higher authority. This makes visionaries dangerous—and often marginalised. Joan was burned at the stake. Ezekiel was mocked and exiled. Even in modern settings, individuals who claim revelatory insight are often stigmatised, medicated, or dismissed. The prophet, it seems, walks a tightrope between genius and heresy, between insight and instability. Yet their visions persist in the collective imagination, precisely because they articulate what cannot be easily said: the inarticulate yearnings of an age, the unconscious anxieties of a culture, the symbolic reconfigurations of individual trauma.

Finally, there is a lingering ethical and epistemological question: *should we trust visionary experience?* Can something neurologically generated still be morally or metaphysically true? The answer may lie not in the origin of the vision, but in its effects. Does it heal, or harm? Illuminate, or obscure? Unite, or divide? The brain may be the medium of revelation, but the measure of its worth lies in the transformation it engenders—both for the individual and their community.

In exploring visions and epileptic trance states, we are not merely probing eccentric corners of mental life. We are examining how the human mind reaches for the transcendent, how it codes emotional truth in symbolic form, and how it constructs narratives that exceed its waking logic. Whether viewed through the lens of neurology, anthropology, or theology, the visionary experience remains a central thread in the fabric of human consciousness. Not a bug in the system—but one of its most enduring features.

The Pineal Gland and Inner Myth-Making Ancient thinkers

Few structures in the human body have been the object of such persistent fascination and metaphysical projection as the pineal gland. This tiny, pinecone-shaped organ—no larger than a grain of rice—sits deep in the centre of the brain, nestled between the two hemispheres near the third ventricle. Anatomically modest, it nevertheless occupies a position of enduring mystery. In classical philosophy, the pineal gland was described by René Descartes as the *seat of the soul*, the singular point at which the immaterial mind and the physical body were thought to interface. In various esoteric traditions, it has been equated with the *third eye*, a gateway to inner vision and mystical insight. Today, it is known primarily for its regulatory role in producing melatonin and maintaining circadian rhythms. Yet these explanations—scientific, spiritual, and symbolic—need not be mutually exclusive. Rather, they map overlapping territories of function, metaphor, and meaning, which together offer a profound window into the mechanisms of imagination, dream, and myth-making.

To begin with the known: the pineal gland is the brain's central timekeeper. It synthesises melatonin from serotonin, primarily in response to light signals received from the retina via the suprachiasmatic nucleus of the hypothalamus. During periods of darkness, melatonin levels rise, inducing sleep and modulating the body's perception of night. Circadian regulation, in this context, is not merely a biological convenience; it is a foundational element of consciousness. By governing the boundary between wakefulness and sleep, the pineal gland indirectly shapes the threshold between external perception and internal imagery. This threshold is critical, for it is where waking cognition blurs into dreams—where logic loosens its grip, and the mind begins to generate the symbolic narratives that populate our inner worlds.

Herein lies the gland's deeper psychological significance. The pineal gland operates not only as a mediator of sleep cycles, but also as a subtle orchestrator of the dream-state. While dreams are produced by distributed networks involving the limbic system, prefrontal cortex, and brainstem, the timing and initiation of the dreaming phase—particularly REM sleep—are regulated in part by the neurochemical influence of melatonin. More than just a sleep hormone, melatonin interacts with GABAergic and serotonergic systems to lower the threshold for visual and emotional imagery. In this sense, the pineal gland does not *cause* dreams, but it sets the stage for them to unfold, acting as a kind of backstage manager of the theatre of the mind.

This theatre is not a neutral space. Dreams are not random noise; they are structured experiences shaped by memory, emotion, and culture. Carl Jung understood this well when he posited the existence of the *collective unconscious*, a reservoir of archetypes and primordial images that surface in dreams and myths across cultures. The recurrence of motifs—labyrinths, serpents, wise elders, shadowy doubles—suggests that the dream-state taps into deep narrative grammars embedded in human cognition. If the pineal gland serves as a switch or conduit into this realm, then it becomes not merely a biological gland but a *cognitive portal*—a threshold organ between physiology and story, between neurochemistry and myth.

It is not difficult to see why ancient thinkers attributed metaphysical importance to the pineal gland. Its anatomical position, tucked away in the centre of the brain and not mirrored bilaterally like most other organs, gave it a unique status. Galen referred to it as the "sphincter of thought," a valve that regulated the flow of psychic pneuma. In later Hermetic and Gnostic traditions, it became identified with the "inner eye," capable of perceiving truths beyond the material. This symbolic association persists in contemporary spiritual circles, where practices such as meditation, rhythmic breathing, and psychedelic ingestion are said to "activate" the pineal gland, granting access to higher consciousness or spiritual revelation. While much of this language is metaphorical or speculative, it reflects a long-standing intuition: that somewhere within the brain lies a structure attuned not merely to time and light, but to meaning.

Modern neuroscience, cautious but increasingly interdisciplinary, has begun to circle back to these intuitions with new tools. Functional MRI studies show that during deep meditation or near-death experiences, the brain exhibits increased synchrony between regions involved in self-perception, memory, and visual processing. Although the pineal gland itself is difficult to image directly due to its size and location, it is implicated in these states via its role in regulating neurochemical states that facilitate altered consciousness. Some researchers have hypothesised that endogenous dimethyltryptamine (DMT), a potent hallucinogen found in trace amounts in the human body, may be produced in or near the pineal gland, especially under extreme physiological stress. While the evidence for this remains inconclusive, it has reignited interest in the gland as a possible mediator of visionary states—those uncanny experiences of vivid imagery, time distortion, and noetic certainty reported by mystics, near-death experiencers, and users of psychedelic substances.

The question is not whether the pineal gland "stores" spiritual insight, but rather whether it plays a *facilitative role* in shifting consciousness into a state where myth-making becomes spontaneous and necessary. When the rational mind quiets, and the barriers of temporal logic dissolve, the psyche begins to speak in image and symbol. In dreams, we do not reason—we *see*. We do not argue—we *narrate*. And it is this narrative, richly metaphorical and emotionally charged, that provides the raw material for the construction of personal and cultural myth. The pineal gland, by modulating the neurochemical conditions under which such images arise, becomes an indirect architect of this mythopoetic capacity.

Importantly, this capacity is not a vestigial remnant of primitive cognition. Myth-making is not opposed to reason; it is *parallel* to it. Where rationality seeks precision and consistency, myth seeks resonance and coherence at the level of meaning. In evolutionary terms, both are adaptive. Logic enables us to manipulate tools and solve immediate problems. Myth enables us to encode collective memory, transmit values, and prepare for the unknown. The pineal gland's role in facilitating the dream-state—and by extension, the symbolic imagination—may be one of the mechanisms through which the brain rehearses moral dilemmas, simulates threats, or integrates traumatic experience. In this way, the gland becomes part of the architecture of psychological resilience.

Consider how often revelations, epiphanies, or creative breakthroughs occur upon waking. These are not random occurrences. During REM sleep, the brain's default mode network—linked to introspection, self-narrative, and memory consolidation—becomes unusually active, while inhibitory networks loosen. This allows for associative leaps and image-based integration that elude waking logic. The pineal gland, by maintaining the circadian integrity of these cycles, indirectly curates the conditions for insight. It does not generate wisdom ex nihilo. But it clears the space for wisdom to emerge.

Where this becomes particularly relevant is in the study of prophets, visionaries, and mystics—individuals whose inner experiences have shaped religions, revolutions, and cultural paradigms. Whether through dreams, trances, or nocturnal visitations, these figures often report encounters with entities or messages that feel simultaneously internal and external, symbolic and specific, ineffable and unforgettable. Modern psychology might explain these as manifestations of the subconscious; neurobiology might point to altered thalamocortical rhythms. Yet neither explanation diminishes their impact. These are not merely neural artifacts—they are *functional myths*, organising forces that transform individual perception into collective meaning.

The pineal gland's significance, then, lies not in its mystical identity per se, but in its *functional centrality to the human experience of meaning.* It reminds us that physiology and imagination are not opposites, but collaborators. That symbolism arises not in spite of biology, but through it. And that somewhere in the brain's shadowed core lies a tiny gland that, by regulating the cycles of sleep and perception, opens the door to inner worlds—where gods are born, dreams are tested, and myth finds its voice.

Dreams as Archetypal Memory Jungian interpretations

The dream-state, often dismissed as a psychological echo chamber or neurochemical by-product, is in fact a medium of transmission. Not merely of repressed wishes or random memory fragments, as Freud once insisted, but of deeply patterned, intergenerational symbols. Carl Jung understood this with uncommon clarity: that the symbols populating our dreams—serpents, floods, caves, birds, masks, and mothers—are not personal inventions but *archetypes* embedded in the psyche. These motifs recur across cultures, epochs, and individuals not because they are learned, but because they are *inherited.* This radical claim—that dreams may carry vestiges of ancestral experience—is not mystical indulgence. It is a serious hypothesis about how the human mind stores, processes, and transmits existential memory across generations.

Jung called this the *collective unconscious*, a psycho-structural inheritance that undergirds all individual experience. Where personal memory is episodic, the collective unconscious is structural; it provides the grammar, not the vocabulary, of inner life. Archetypes, in Jung's system, are not images but *tendencies to form images*—primordial organising patterns that take symbolic shape under the pressure of emotion, crisis, or transformation. Thus, when a modern dreamer sees a great flood swallowing the world, or a monstrous beast emerging from the shadows, it is not necessarily autobiography. It may be *deep memory* surfacing, encoded not in neurons but in narratives.

This raises a provocative question: if archetypes are universal, could some be rooted in *actual events*—traumas so foundational they left psychic residues? Consider the flood myth, which appears in Sumerian, Babylonian, Biblical, and Mesoamerican traditions. These are not identical stories, but they converge on a symbolic structure: water as chaos, deluge as punishment, and survival through select preservation. Geological evidence confirms that the end of the last Ice Age brought catastrophic flooding to many inhabited regions. The Black Sea deluge hypothesis, for instance, suggests a sudden inundation around 5600 BCE when Mediterranean waters breached into the freshwater basin. Such a cataclysm, witnessed and mythologised by early agricultural societies, could have seeded a narrative that mutated through oral transmission into enduring symbolic form. When such a motif recurs in dreams millennia later, it may not be metaphor alone—it may be *mnemonic sediment* from ecological trauma.

Dreams, after all, are not merely expressions; they are simulations. Cognitive neuroscientist Antti Revonsuo has argued that dreaming evolved as a threat-simulation system, enabling early humans to rehearse survival scenarios. But what if those simulations are not just rehearsals but *replays*—reflexive reconstructions of ancient crises? Consider the ubiquitous dream of being chased, often by a vague but terrifying pursuer. In clinical settings, this motif frequently appears in patients with anxiety, PTSD, or unresolved attachment trauma. Yet it also mirrors ancient predator-prey dynamics that shaped hominin evolution. Our ancestors spent thousands of generations in precarious proximity to big cats, snakes, and rival hominids. Is it so far-fetched to imagine that dream content would retain traces of those encounters—rendered not as memory per se, but as encoded vigilance?

This perspective reframes archetypal dreams not as myth-making in the poetic sense, but as *cognitive inheritance*—mental structures shaped by evolution, refined by culture, and sustained through storytelling. It also reorients Jungian theory away from abstraction and toward empirical footing. Just as we inherit fear of heights or aversion to rotting smells, we may inherit symbolic repertoires that once carried adaptive value. The "great mother," the "wise old man," the "hero's journey"—these are not just literary tropes but psychic scaffolds for navigating developmental stages, social roles, and existential threats. They recur in dreams because they are not learned anew each generation; they are *activated.*

The implications of this model are wide-ranging. In psychotherapy, archetypal dreamwork enables patients to locate personal experiences within a larger symbolic framework, reducing alienation and enhancing narrative coherence. In anthropology, it opens the door to interpreting myths not only as cultural artefacts but as neuropsychological fossils. In evolutionary psychology, it suggests that dreams are not vestigial mental noise but *functional simulations* rooted in ancestral adaptation. And in trauma studies, it underscores how repeated exposure to collective stress—whether war, migration, or ecological collapse—can imprint itself not just on social memory, but on the *architecture of dream itself.*

Migration is a particularly potent context for this. Human history is punctuated by waves of forced movement—due to climate change, famine, conflict, or conquest. Each migration involves rupture: from land, from lineage, from linguistic stability. And these ruptures often resurface in dreams. Studies of second-generation refugees and diasporic communities show elevated incidences of displacement motifs in dreams: lost homes, collapsing borders, forgotten languages. These are not simply reflections of current anxiety; they are *intergenerational echoes*—the psyche's attempt to metabolise historical dislocation. In such cases, the dream becomes a bridge between generations, a symbolic translation of trauma that speech alone cannot fully encode.

Ecological catastrophe adds another layer. From the Younger Dryas cold snap to the Bronze Age droughts to the Dust Bowl of the 1930s, humans have repeatedly encountered the limits of environmental predictability. These encounters do not vanish from consciousness once the crisis abates; they become mythologised. Consider the prominence of desolate landscapes, burning skies, or dead animals in dreams. While some may be autobiographical or culturally primed, others may tap into *species-level memory* of ecological threat. The contemporary resurgence of apocalypse dreams in the context of climate change—particularly among young people—is not just social anxiety. It may represent a kind of *archetypal reactivation* of older, buried patterns of collapse and adaptation.

Of course, not all dreams bear this ancestral weight. Many are personal, idiosyncratic, or fragmentary. But the *structural motifs*—labyrinths, shadow figures, primal animals, celestial omens—persist with startling consistency. The question is not whether they are "true" in the empirical sense, but whether they are *valid* as symbolic forms through which the psyche processes complexity. Jung's brilliance lay in recognising that meaning does not reside only in clarity or correspondence. Meaning can also be found in resonance, in repetition, in the shape of a symbol that returns again and again across the sleep of centuries.

Critics have long challenged Jung for his vagueness or for the Eurocentric bias of some of his categories. And indeed, any theory of archetypes must grapple with cultural specificity. The serpent may mean wisdom in one context and evil in another; the cave may symbolise rebirth, retreat, or death depending on the narrative frame. Yet even these divergences underscore the deeper point: archetypes are not fixed meanings but *structural attractors*—patterns of salience that cultures interpret differently but gravitate toward again and again. To study archetypal dreams is not to impose a single mythology, but to map the symbolic grammars by which different peoples encode common human dilemmas.

If the dream is a text, then archetypes are its syntax. But the language of dreams is not written in words—it is composed of image, affect, and sequence. This is why dream interpretation cannot be mechanised or reduced to formulae. It requires attunement, empathy, and an appreciation for ambiguity. A dream of drowning may signal repression for one person, ecological anxiety for another, and ancestral memory of deluge for a third. Yet the recurrence of water as a threshold symbol—between life and death, known and unknown—suggests a deep structure that transcends individual context.

Ultimately, Jung's theory of archetypal memory challenges the modern tendency to compartmentalise mind and history, biology and myth. It offers a model in which the dream becomes a meeting place between the neuronal and the numinous. Between the story of a person and the story of a species. In this view, our dreams are not ours alone. They are *inheritances*—symbolic strands in a much older weave of memory, imagination, and survival. We do not merely dream; we are *dreamed* by something older than self, older than culture—something that remembers the fire, the flood, the beast in the tall grass, and the long road home.

The Oracle Brain Could some prophets

The concept of prophecy, once the exclusive domain of religion and myth, is increasingly open to neurological scrutiny. Ancient cultures revered certain individuals for their visionary powers—prophets, sibyls, oracles, seers. They were consulted in times of crisis, feared during moments of political unrest, and sometimes persecuted when their insight veered too far from orthodoxy. But what if these individuals were not mere storytellers or spiritual figureheads? What if they possessed a neurocognitive profile that made them *genuinely different*—a kind of cognitive divergence characterised by heightened pattern recognition, deep temporal intuition, and hyperconnectivity between otherwise distinct brain regions? The "oracle brain," in this hypothesis, is not a metaphor. It may be a neurological reality—rare but real, and evolutionarily significant.

The idea that prophets might exhibit anomalous but functional brain patterns finds early support in anthropology. Across cultures, shamans and visionaries often emerge from the social periphery—those deemed "eccentric," "touched," or "mad." Yet their insights were indispensable. They detected omens, interpreted dreams, navigated uncertainty. Unlike priests, whose roles were institutional and doctrinal, prophets were spontaneous nodes of *pattern recognition*. They intuited connections between natural signs, political shifts, and internal visions. This behaviour aligns closely with what cognitive scientists today describe as *apophenia*—the tendency to perceive meaningful patterns in noise. But rather than dismiss this as delusion, we must ask: when is apophenia *adaptive*?

Cognitive divergence is increasingly understood not as dysfunction but as alternative optimisation. Savant syndrome, synaesthesia, temporal lobe epilepsy, and autism spectrum profiles all reveal ways the human brain can reconfigure its connectivity in ways that produce non-normative—but sometimes highly valuable—perceptions. The prophetic mind may represent a clustering of such traits: elevated connectivity between default mode and salience networks; increased activity in right temporal and parietal regions; sensitivity to interoceptive signals; and a low perceptual threshold for anomaly detection. These neural attributes, while they may manifest as social estrangement or verbal idiosyncrasy, can also result in visionary synthesis—especially when placed within a ritual, symbolic, or communal framework.

Take the example of Ezekiel's vision in the Hebrew Bible—a fiery chariot, wheels within wheels, angelic beings flashing like lightning. To modern readers, it reads as either allegory or hallucination. Yet recent neurological studies suggest that similar experiences are reported by individuals with temporal lobe epilepsy or limbic system hyperactivity. Such episodes often involve intense visual imagery, a feeling of numinous presence, and altered perceptions of time and space. Joan of Arc, whose auditory visions guided military strategy and political prophecy, would today likely be evaluated for hallucinatory or schizotypal traits. But in their respective historical contexts, such neurodivergence was not pathological—it was *revered*. Their brains, in a sense, operated on a frequency that detected and synthesised signals—psychological, ecological, sociopolitical—that others missed or ignored.

Modern neuroscience supports the plausibility of such a brain. Individuals with *increased cross-talk between hemispheres*—such as in cases of atypical corpus callosum architecture or enhanced interhemispheric coherence—often exhibit intuitive leaps, metaphorical thinking, and nonlinear association patterns. These are hallmarks of mythopoetic insight, a trait commonly reported in prophetic figures. Moreover, dopamine and norepinephrine levels—especially their modulation of salience attribution—play a significant role. Too little, and perception flattens; too much, and the world brims with overmeaning. A delicately calibrated oracle brain may occupy a space just shy of pathology—teetering at the edge of delusion, yet still grounded enough to articulate insights that resonate.

Importantly, this is not a romanticisation of mental illness. The oracle brain is not defined by suffering, but by *sensitivity*. It does not require dysfunction, but *difference*. And unlike clinical diagnoses that seek to reduce symptoms, the prophetic condition—if it exists—was traditionally *integrated* into culture. Rituals, incubation chambers, dietary constraints, trance dances, and dream incubation were not arbitrary spiritual practices; they were *technologies* for stabilising and interpreting the insights of these cognitively divergent individuals. The Oracle at Delphi, for instance, entered altered states induced by subterranean vapours, rhythmic chanting, and ritual preparation. Her utterances, though cryptic, were taken seriously because they were encoded within a cultural framework that *valued anomaly*.

Contrast this with modern contexts, where divergent cognition is often medicalised or marginalised. Pattern-hungry minds are categorised as paranoid, obsessive, or psychotic. But perhaps this says more about the narrow bandwidth of modern epistemology than about the minds themselves. Pattern recognition—especially in uncertain environments—is a powerful adaptive trait. Predicting the weather from cloud formation, the success of a harvest from animal behaviour, or the onset of conflict from subtle social cues—all require high sensitivity to pattern, context, and deviation. The prophetic brain may be an evolutionary artefact of this adaptive niche: the human node for uncertainty navigation.

From a neuroanthropological perspective, prophecy may have emerged not from divine intervention, but from *cognitive specialisation*. Every tribe or early civilisation needed its strategist, its visionary, its synthesist of the unseen. These individuals did not conform to consensus reality. They operated from the margins—neither fully rational nor irrational, but *transrational*. They saw connections others dismissed. They warned of catastrophe, proposed new ethical frameworks, reoriented collective purpose. That their visions were encoded in religious or mythic language does not invalidate their function; it contextualises it. Myth, after all, was the cognitive container through which extraordinary perception became socially digestible.

Some of today's most compelling thinkers—systems theorists, futurists, AI ethicists, environmental prophets—may carry echoes of this ancestral role. Their insights often arise from synthesising disparate domains, perceiving patterns across time and space, and tolerating ambiguity long enough for novel structure to emerge. Like ancient prophets, they are often dismissed at first, only to be heeded in hindsight. Their brains, too, may be wired differently—not impaired, but *hyperconnected.* The oracle brain, in this context, is not a historical relic but an ongoing variant within the neurodiversity of our species.

This reframing carries ethical implications. Instead of dismissing visionary perception as esoteric or pathological, we should ask what conditions *amplify or suppress* this mode of insight. Educational systems, pharmacological interventions, and cultural norms all shape the expression—or repression—of divergent cognition. The current emphasis on speed, productivity, and narrow rationality may be stifling a mode of thought essential for long-range thinking and deep pattern recognition. If we value foresight, creativity, and synthesis, then our cultures must cultivate—not merely tolerate—cognitive deviance.

Moreover, we must learn to discern between delusion and insight without reducing the latter to empirical data alone. Prophetic thinking often manifests *before* the evidence is in—through intuition, symbol, or radical synthesis. It does not replace science; it precedes it. Many major breakthroughs in science, ethics, and art began as *visionary hypotheses*, scorned in their time. The role of the oracle, then, is not to predict events like a clock forecasts the hour. It is to *perceive structure*—emerging complexity, moral dissonance, systemic imbalance—and give it symbolic shape.

The oracle brain may be rare, but it is not extinct. It may even be more necessary now than ever. As civilisation accelerates, interconnects, and destabilises, the ability to *see across systems*, to feel emerging tensions before they surface, and to communicate them with narrative potency will become a core human need. Whether these insights emerge from artists, scientists, or spiritual practitioners is secondary. What matters is that we recognise the *neurodivergent gift* at the heart of human foresight—and build cultures that can hear it.

We once burned prophets or built temples around them. We now tend to ignore or medicate them. A wiser society might do neither. It might instead listen carefully, test respectfully, and protect the fragile brilliance of minds that—while different—are not deficient. For in the intricate circuitry of the oracle brain lies a capacity we cannot afford to lose: the capacity to make meaning, warn of danger, and imagine the shape of things to come.

Chapter 8
The Language Before Language — Sound, Symbol, and Sacred Geometry

Language, as we encounter it today, is hardly the genesis of human expression. It is a refined façade—evolved, cultivated, and tamed—resting atop something older, deeper: patterned sound, symbolic geometry, and a kind of non-verbal logic embedded in myth and ritual. Long before alphabets or syntax, before the subject–object split that underpins sentence structure, human societies operated in realms of meaning that were lived rather than merely articulated. This chapter turns towards that hidden substratum—the language before language—not as whimsical conjecture, but as an interdisciplinary investigation into how consciousness has, for millennia, encoded, transmitted, and safeguarded knowledge through channels more elemental than grammar.

This is not an exercise in mystical nostalgia, nor a bid to drape language in gauzy spiritual sentiment. It is instead a call to examine the neurological, cognitive, and mathematical foundations of human communication through the lens of mythopoetic intelligence—those primal ways of knowing that precede the alphabet. Consider for a moment: might the chants of shamans, the sacred geometries chiselled into temple walls, the archetypal symbols carved in stone, and the so-called 'angelic languages' of mystics have been more than aesthetic or devotional flourishes? Could they have functioned as precise cognitive tools—mechanisms for storing knowledge, modulating awareness, and triggering deeply wired recognition patterns in the human mind?

One entry point lies in phonetic vibration. Across continents and ages, certain syllables—OM in India, HU in Sufi practice, AH in various meditative traditions—are far more than religious ornaments. They pulse at frequencies that shape breathing, sharpen attention, and alter bodily state. These sounds are not the product of chance. They mirror neurological resonances, perhaps even inherited memory from earlier stages of our evolution. Ancient deities are often named in patterns of sound—elongated vowels, resonant nasals, abrupt glottal stops—that stir the listener physically, bypassing rational interpretation altogether. Such phonemes exist before meaning; they act directly on flesh and nerve, much as the contours of a landscape predate any map.

From sound to form, we find sacred geometry—not as arcane decoration, but as a mental scaffold. Spirals, mandalas, hexagrams, labyrinths: these are structural languages in their own right. They map motion, energy, and time itself. The Fibonacci sequence, the golden ratio, the endlessly repeating rhythm of fractals—these are not merely modern discoveries; they are etched in Neolithic rock art, echoed in megalithic architecture. Such shapes are not idle embellishment; they narrate. The hero's journey traces a spiral; the cycle of seasons folds back into itself like a torus; the mandala holds both a cosmological map and a psychological mirror. Ancient storytelling, then, was not confined to sound—it was spatial, sculpted in stone, rhythm, and proportion. Geometry was not the ornament of myth; it was its skeleton.

Symbols, too, carry more weight than ornamental beauty. They are devices of compression—mental archives reduced to a single, potent sign. The ankh, the swastika in its sacred, pre-political form, the ouroboros: these are condensed cosmologies. They operate more like a compressed digital file than a corporate logo, each holding within it a network of meanings that unlock only when the mind engages with its corresponding myth or ritual. The recurrence of such motifs in unrelated civilisations—a phenomenon Jung sought to explain with archetypes—hints not only at diffusion, but at convergent cognition. These forms may spring from shared neurological wiring, surfacing again and again as visual echoes of universal human experiences: life and death, order and chaos, fertility and danger, transcendence and return.

The most elusive strand in this web is that of "lost languages": Enochian, the language of the birds, the Adamic tongue. Modern linguistics often sets them aside as deliberate invention, ecstatic babble, or theatrical mysticism. Yet they deserve a second look—not as literal proto-languages, but as crafted exercises in reconfiguring thought. The Enochian system devised by John Dee and Edward Kelley in the sixteenth century, for example, is strikingly systematic, governed by an internal logic and supported by geometric 'tablets' that carry a symbolic grammar. It behaves almost like a programming language for the mind, its syntax tuned to induce altered states. Such constructed tongues may function as laboratories for pre-verbal thought, drawing the mind into nonlinear connections, mental imagery, and sensory crossovers between sight and sound.

The proposition here is that what we now recognise as language arose from a more ancient, integrated network of symbolic patterning—where sound, shape, and image formed a seamless semantic web. Before words, there was resonance; before script, symmetry; before narrative prose, a ritual syntax woven from patterns of perception. In this view, myth was never mere entertainment or explanation—it was a living archive, preserving the scaffolding of emotional, cosmological, and ecological knowledge across centuries.

Such a perspective alters how we approach the relics of early culture. A Babylonian cylinder seal may hold not simply imagery but frozen syntax. A Vedic hymn can be read as a sonic equation. Petroglyphs might record patterns of recurrence, not isolated events. The mnemonic feats of oral cultures—from Aboriginal songlines to the Homeric epics—become all the more impressive when we grasp that they compress environmental, astronomical, and moral knowledge into rhythm and metaphor.

To speak of a "language before language" is to dig beneath the cultural interface into the operating system itself. It demands that we set aside our bias towards literacy—that reflexive belief that meaning begins with writing. It asks us to entertain the possibility that much of what we file under religion, art, or mythology once formed part of a single semiotic toolkit—finely tuned for embodied thought, generational memory, and heightened states of awareness.

In an age when attention is fragmented, memory outsourced, and our symbolic repertoire increasingly thinned, the erosion of this older language is more than a scholarly curiosity—it is an existential problem. Reconnecting with it does not mean turning back the clock to a primitive state; it means restoring the full spectrum of our cognitive capacities. This chapter considers how the ancient interplay of sound, symbol, and form might broaden our understanding of language, intelligence, and memory—and how the long-buried grammars of myth could yet influence the future shape of meaning.

The Primal Sounds of Myth Investigates

The most ancient stories were not written—they were sung, chanted, and intoned. Long before myth migrated to ink and papyrus, it travelled on breath and vibration, through mouths and memory. It is a mistake to regard early myth as merely linguistic content wrapped in sound; rather, the sound *was* the content. In primal cultures, the mythic utterance—whether the name of a god, a sacred syllable, or a repeated refrain—was not simply a descriptor or narrative device, but a functional *technology of attention*, designed to transform consciousness, encode memory, and imprint meaning at a level deeper than semantics. To study the primal sounds of myth is to interrogate the neurology of belief itself.

Anthropological records and contemporary fieldwork in indigenous cultures consistently show that sacred utterance is not treated as metaphorical. It is *operative.* In the Aboriginal Dreamtime traditions, for instance, spoken songlines are believed to create and maintain the land itself; they are both map and medium. In Vedic India, the recitation of Sanskrit mantras like *Om Bhur Bhuvaḥ Swaḥ* is not merely devotional but vibrationally causal. The sound is thought to *do* something—to the environment, to the body, to the invisible structure of being. What unites these examples is the conviction that certain phonetic combinations carry force beyond the symbolic, rooted in something fundamentally acoustic and physiological.

Modern neuroscience gives partial credence to this ancient intuition. Studies in neurolinguistics and psycholinguistics have revealed that certain phonemes—particularly nasalised vowels, hard plosives, and sibilants—activate predictable regions of the brain associated with attention, memory consolidation, and emotional salience. The phenomenon of "glossolalia" or speaking in tongues—often dismissed as religious hysteria—has been shown to reduce activity in the frontal lobe (responsible for executive control) while increasing activity in areas associated with self-perception and language rhythm. These states closely resemble trance and meditative absorption. In this sense, repeated vocalisation of certain sacred sounds may function as a neurological lever—suspending the default mode network, sharpening sensory focus, and opening access to symbolic or visionary content.

The names of gods, too, show remarkable phonetic consistency across disconnected cultures—particularly in their use of open vowels and bilabial or glottal consonants. Consider the following: *El*, *Alah*, *Amun*, *Anu*, *Odin*, *Yahweh*, *Ra*, *Ba'al*, *Iaou*, *Om*, *Ahura*, *Enki*, *Isis*, *Astarte*. These names are dominated by sonorous vowels, frequently framed by aspirants or resonants. This is not accidental. Linguist John Ohala's "frequency code" hypothesis suggests that lower-frequency sounds are neurologically processed as more dominant or authoritative, while higher-pitched phonemes tend to denote vulnerability or submissiveness. The reverence triggered by divine names may have less to do with theological meaning than with acoustic affect—a direct stimulation of primal dominance-submission circuits in the mammalian brain.

Such effects are not confined to names. Whole myths may have been designed as *auditory rituals*, structured to induce particular states of consciousness. The use of rhythmic chanting, alliteration, internal rhyme, and cyclical phrasing are common features of sacred narratives globally. The Homeric epics, for instance, are not just long poems but mnemonic matrices, encoded with formulaic phrases ("rosy-fingered dawn," "wine-dark sea") that served both as aesthetic devices and memory anchors for oral transmission. Vedic hymns follow rigorous metrical and phonetic patterns—*chandas*—that are not simply poetic traditions but techniques of vibrational consistency. This suggests that early myth was not composed but *tuned*, like a sonic mechanism calibrated to the psyche of the listener.

The question arises, then: why did these structures work so consistently across time and culture? The answer may lie in the overlap between phonetics, emotion, and memory. Brain imaging studies show that emotionally charged words—particularly those associated with sacred or taboo content—activate the amygdala and hippocampus more intensely than neutral language. When those words are repeated rhythmically or paired with breath control (as in chanting), their imprint deepens. The experience of 'feeling the truth' of a mythic story may arise not from intellectual assent but from *neurological entrainment.* This makes primal sound not an aesthetic flourish of early culture, but its primary epistemological tool.

This perspective disrupts the traditional literary or theological framing of myth. It shifts the axis of meaning from content to *resonance.* Myths are not only stories about gods and heroes—they are sonic environments that organise human attention and trigger deep associative networks. In this view, the mythic experience is not conveyed; it is *instantiated* in the body of the listener. Sound becomes the medium through which cultural memory and cosmological orientation are internalised.

There is also reason to believe that some of these primal sound patterns were embedded in sacred architecture. Ancient temples and ceremonial chambers often exhibit anomalous acoustic properties. The Hypogeum of Ħal Saflieni in Malta, for example, amplifies low-frequency male voices in a way that causes auditory hallucinations and somatic vibration in the listener. The King's Chamber of the Great Pyramid of Giza has a resonant frequency that amplifies tones in the 90 Hz range—precisely the frequency shown to produce meditative states in lab settings. Such spaces suggest that *acoustic engineering* was an integral part of ancient sacred experience. The sound was not incidental; it was *the point.*

Critically, this is not to suggest that all ancient people were scientists in the modern sense, nor that they uniformly understood these effects at a theoretical level. But they were, in many cases, masterful empiricists of human consciousness. Through centuries of oral experimentation, ritual iteration, and cultural selection, they discovered what worked—and encoded it into their practices, stories, and names. The 'truth' of myth, then, may not lie in its propositional content but in its capacity to *modulate neural states and encode cultural intelligence in sound.*

Of course, the risk of romanticising these traditions is real. Not all ancient chants are meaningful, and not all myths were designed with acoustic precision. But the consistent cross-cultural recurrence of specific phonetic strategies, symbolic vocalisations, and resonant narrative patterns suggests a more foundational role for sound than is generally acknowledged. It implies that what we now consider "meaning" may have begun as *resonant pattern recognition*—a somatic intelligence tuned to vibratory symmetry long before it developed abstract logic.

This carries implications beyond myth. In contemporary environments increasingly dominated by visual stimuli and textual information, we risk losing the depth of auditory intelligence that shaped our evolutionary and cultural past. There is evidence that certain phonetic environments can improve retention, regulate mood, and sharpen attention. The growing popularity of binaural beats, sound baths, and vocal toning in wellness culture reflects, albeit crudely, a rediscovery of something primal: that sound, particularly rhythmic, harmonically structured sound, is not ancillary to cognition—it *is cognition* in one of its oldest forms.

This subchapter invites us to re-examine myth not as allegory or metaphor, but as a complex, neurologically active sound system. It suggests that our ancestors did not merely speak myths—they *sounded them into being*, tuning language, breath, and body into alignment with memory and meaning. What we inherit today in script and text may be only the shadow of a deeper, vibrational literacy—one that resonated not in the eye, but in the nervous system, in the flesh, and in the sacred silence that follows the last syllable of a name intoned just right.

"Sacred Geometry as Narrative DNA"

A myth is not merely a story. It is an architecture. Beneath the surface of heroes, monsters, and cosmic battles lie invisible geometries—spirals of rebirth, ratios of harmony, nested cycles of return—that mirror the mathematical scaffolding of the natural world. In cultures separated by oceans and epochs, we find tales shaped not only by language and belief, but by structure—stories that unfold in ways uncannily aligned with principles of sacred geometry. The implication is not mystical but methodological: just as DNA encodes biological form through molecular symmetry, myth may encode cultural and psychological form through geometric resonance.

When early mythmakers traced the hero's path through descent, ordeal, and return, they were not simply narrating events—they were inscribing arcs that echoed the golden spiral of a nautilus shell or the harmonic ratios of musical scales. In mythic structure, there is often a recursive elegance—patterns within patterns, stories nested in cycles, each part reflecting the whole. Consider the archetype of the "world tree," found in Norse Yggdrasil, the Mayan Ceiba, and the Vedic Ashvattha. These are not just trees—they are vertical cosmograms, uniting heaven, earth, and underworld. Their form is fractal: trunk to branch to twig to leaf, the same pattern replicated at multiple scales. The story they embody—of descent and ascent, of branching life and returning source—mirrors the recursive mathematics of natural systems.

Sacred geometry refers not to aesthetic ornamentation but to mathematical principles observable in nature: the Fibonacci sequence, the golden ratio (phi), the hexagonal symmetry of honeycombs, the fractal recursion of ferns. When these patterns surface in myth, they often do so unconsciously, embedded in the very logic of the narrative. The "threefold" nature of many mythic quests—the beginning, middle, and return—is not merely a literary convention. It reflects the tripartite division of space and time found in physical systems: entropy, equilibrium, and renewal. Joseph Campbell's monomyth, or Hero's Journey, mirrors a toroidal cycle—a closed-loop flow with a central inversion point, akin to the energetic topology of black holes, magnets, and even ecosystems. It is not surprising that these same tropes recur in cultures with no historical contact. The brain, conditioned by nature's geometry, naturally reproduces these forms in narrative.

Ancient architecture often reveals an intimate understanding of this relationship between myth and form. The floorplans of Hindu temples, for example, are based on mandalas—geometric diagrams that represent cosmological principles. The Sri Yantra, a complex sacred figure composed of interlocking triangles, is not simply symbolic; it is a schematic representation of the union of opposites, of feminine and masculine forces, of spiritual progression toward a central stillness. Likewise, the Great Pyramid of Giza encodes proportions within a near-perfect golden ratio, suggesting an alignment of mythic intention (the journey to the afterlife) with mathematical precision. These structures did not merely house mythic rituals—they embodied them. Sacred space was not just spatial but informational: it was geometry imbued with narrative purpose.

Modern mathematical physics offers unintentional confirmation of this convergence. Fractals—self-similar structures that repeat across scales—are now recognised as central to the modelling of coastlines, vascular systems, neural networks, and even market behaviours. The brain itself is organised fractally, with dendritic structures that mirror trees and lightning. This fractal architecture allows for both compression and expansion of information, much like oral myth: a short phrase—"As above, so below"—can recursively encode entire cosmological systems. The same economy is found in religious iconography. A yantra, a Celtic knot, or a Mayan glyph can communicate volumes without words, provided the viewer knows the underlying structure. These are not symbols in the modern sense; they are *cognitive geometries*, compact codes designed for repetition, transmission, and resonance.

There is also a neurological dimension to the appeal of geometric structure in myth. Studies in neuroaesthetics suggest that the human brain exhibits preferential responses to symmetry, proportion, and harmonic balance. Exposure to Fibonacci sequences and golden ratio proportions has been shown to activate reward circuits in the ventral striatum and orbitofrontal cortex. These are the same circuits implicated in aesthetic pleasure, goal-directed behaviour, and even spiritual experience. It stands to reason, then, that myths encoded with these geometric ratios would *feel true* at a level deeper than rational analysis. They would evoke a sense of rightness, inevitability, or profound insight—not because of what they say, but because of how they are shaped.

This is perhaps why myths endure not through explanation, but through repetition. A myth told in sacred space, aligned with the seasons, mirrored in the architecture of ritual and chant, becomes not just a story but a *resonant system*—a pattern that enters the body through multiple channels. The geometry of the narrative is mirrored by the geometry of the place, the movement, and the sound. In such contexts, myth functions less like fiction and more like *algorithmic inheritance*—a way of storing and transmitting essential information about survival, morality, and cosmology, all encoded in form.

Contrasting views, of course, caution against over-interpretation. Some scholars argue that the recurrence of geometric forms in myth is a byproduct of cognitive constraints—our brains are wired to see patterns, and nature offers only so many. Others note that geometry can be retrofitted onto nearly any narrative, if one is inclined to search for it. These critiques are valid to a point. But they overlook the possibility that form itself is *a kind of meaning*. That stories shaped like spirals, cycles, or fractals are not metaphorical at all—but are themselves the content, encoding information about how change occurs, how systems adapt, and how life repeats and innovates simultaneously.

To treat sacred geometry as narrative DNA is not to claim that myths are blueprints in the biological sense. It is to suggest that mythic stories may replicate the logic of natural design: modular, adaptive, symmetrical, and recursive. This has practical implications. It suggests that myths, like genes, are subject to selection—not for factual accuracy, but for functional resonance. A story that fits the cognitive and emotional architecture of the human brain is more likely to survive, spread, and return in new forms. The geometry of myth is not decorative—it is *evolutionary*.

This is especially relevant in a digital age where information is often stripped of structure and depth. Understanding the geometric intelligence of myth offers a corrective. It reminds us that form matters. That the shape of a story—its rhythm, its symmetry, its cycles—is as important as its content. That deep learning is not linear but patterned, recursive, and dimensional. Sacred geometry, in this view, is not a mystical sidebar to culture. It is the logic of coherence itself—woven into the very stories that teach us who we are, where we come from, and how to return.

As modern science converges with ancient intuition, we begin to see that the boundary between narrative and nature is thinner than once believed. The mythic mind did not need calculus to intuit the golden ratio; it found it in shells, storms, and seeds—and replicated it in stories. What endures across time is not just the tale, but the *structure of the tale*, carrying with it the deeper grammar of life's unfolding. To recognise that structure is to begin decoding the intelligence of our oldest narratives—not as fantasies, but as formal truths, hidden in plain sight.

Symbol as Time Capsule Deconstructs the idea

Throughout human history, symbols have functioned not simply as marks of culture or emblems of identity, but as vehicles for concentrated meaning—semantic payloads engineered to survive time, translation, and even the loss of civilisations. Far from being static decorations or aesthetic flourishes, many of the world's oldest symbols are better understood as *compressible cognitive code*: ultra-compact messages encoded for long-term transmission across generations, geographies, and even epistemological shifts. In this sense, a symbol is less a fixed representation and more a *time capsule*—a mnemonic device designed to preserve knowledge, ritual, cosmology, or ethics in a form that is durable, repeatable, and resilient to erosion by time or dogma.

Consider the spiral—a motif carved into megalithic stone in Malta, Ireland, Peru, and the American Southwest. While its contexts vary, its recurrence suggests that this shape encoded more than artistic whim. In multiple traditions, it signified birth, cycles, death, and return. But what makes the spiral enduring is not just what it means, but how it means. It visually embodies motion and recursion, growth and return. The spiral becomes a visual grammar for change over time, a symbolic distillation of dynamic systems—from the growth of shells to the whirl of galaxies. It preserves not a sentence, but a system. It is a diagram of process, not merely a stand-in for a word. Such symbols are *form-meaning hybrids*, which is precisely why they outlast the languages that name them.

To treat symbols as data containers—shaped for cognition—is not a modern imposition, but increasingly an archaeological and cognitive consensus. The famous "Flower of Life" motif, for instance, found in Egyptian temples and replicated in the art of medieval Jewish mystics and Islamic geometric design, contains within its interlocking circles multiple mathematical insights: hexagonal symmetry, the basis for the Platonic solids, and hints of the golden ratio. These are not incidental; they reflect an intentional encoding of universal principles. That the same symbol appears across radically different epochs suggests that the form itself carries a type of epistemic weight—a container for truths that remain intelligible across the rise and fall of entire civilisations. Symbols such as these are not static artefacts; they are *living algorithms*, offering those who engage them a means to reconstruct knowledge without text.

This use of form as a container for meaning may find its neurological basis in how the brain handles visual information. Humans are inherently symbolic processors: our cognitive systems are predisposed to extract patterns from limited input and assign them layers of contextual meaning. In other words, our brains *compress and decompress* symbols naturally. What begins as a carved circle or a cross becomes, over time and exposure, a matrix of meaning—layered with myth, ritual, cosmology, and instruction. The early Christian ichthys (fish symbol), for example, was not merely a clandestine marker but a *portable theology*—representing Christ, baptism, the feeding of multitudes, and the Greek acronym for "Jesus Christ, God's Son, Saviour." All of this latent content was compressed into a two-stroke image—a time-resistant, text-free transmission of a doctrinal worldview.

The Rosetta Stone is often hailed as a triumph of linguistic decoding, yet it is also a reminder of how symbols—especially hieroglyphs—functioned not as phonetic syllables alone, but as *ideographic encapsulations*. Egyptian hieroglyphs blended sounds, objects, and ideas into single glyphs, capable of communicating across levels: literal, metaphorical, sacred. The Ankh symbol, for instance, is frequently translated as "life," but this obscures its function as a philosophical compression: it evokes breath, divine essence, the flow of water, and the continuity of existence, depending on the interpretive frame. This multidimensionality is not a design flaw—it is the defining feature. Such symbols behave like *semantic holograms*: no matter how they are fragmented, some version of the whole remains embedded within.

Symbols that endure also exhibit qualities similar to what we now call *information entropy resistance*. The less a symbol depends on verbal context, the more likely it is to survive interpretive drift. This is why visual motifs—like the triskelion, the ouroboros, or the labyrinth—maintain their integrity despite cultural upheaval. They are less vulnerable to censorship, translation error, or theological revision than prose or poetry. They require no formal schooling to engage, and yet they reward deeper inquiry with more complex meaning. This is remarkably close to the logic of compression used in modern data systems: maximising information density while minimising susceptibility to noise. In this way, a stone glyph from 3,000 BCE and a QR code from today are conceptually cousins—each one an engineered strategy to preserve meaning across contexts.

But there is also a psycho-cultural layer to this phenomenon. Symbols function not merely as conveyors of explicit information, but as anchoring points in the collective unconscious. Carl Jung famously proposed that symbols emerge from archetypes—primordial images that structure the psyche. Whether one accepts this Jungian metaphysics in full or not, there is strong evidence that humans respond to symbols with emotion-laden immediacy. A swastika—originally a Vedic symbol of auspiciousness—now evokes global moral revulsion due to its appropriation by Nazism. The symbol's *emotional charge* has been overwritten, but its power has not diminished. This ability of symbols to retain and transmit emotional valence across time and space suggests they are not merely representational tools but *affective containers*, capable of preserving cultural trauma or reverence for generations.

The power of symbols also lies in their ability to operate simultaneously at different cognitive altitudes. A child sees the crescent moon and star and may think only of the night sky. A Muslim sees the same image and recalls the lunar calendar, the rhythms of prayer, and centuries of civilisational identity. A historian sees Byzantine origins. A mystic sees divine unity. One symbol, multiple payloads. The cognitive elegance of this system is that it allows for continuity without requiring full ideological consensus. Unlike text, which demands linear decoding, symbols invite layered interpretation. They are *modular containers*, open to new meanings while preserving old ones in dormant form. This design principle is not accidental; it is a survival mechanism for meaning itself.

It is not only ancient cultures that understood this principle. Modern iconography—from corporate logos to protest graffiti—continues to exploit the time-capsule function of symbols. The Apple logo, for instance, is more than branding; it is a compressed myth of innovation, rebellion, knowledge (the biblical fruit), and sleek design. National flags, religious insignias, and even traffic signs function with similar economy: maximum recognition, minimal explanation. The modern world is flooded with symbols, but we have largely forgotten that they can encode *wisdom*, not just information. In rediscovering the symbolic literacy of our ancestors, we may regain tools for long-term thinking in an age obsessed with immediacy.

There is also a philosophical proposition embedded here: if a culture wishes to speak to the future—to descendants who may no longer speak its language or share its worldview—symbols may be its most effective medium. This is the rationale behind projects like the "nuclear semiotics" movement, which seeks to warn future civilisations of radioactive waste sites. Linguists and artists have proposed symbols that evoke unease, decay, or taboo without relying on any known tongue. The challenge is staggering: how do you design a message to be legible 10,000 years from now? And yet ancient cultures did precisely this, whether intentionally or not. Their myths, rituals, and carved signs remain partially legible to us across vast gulfs of time because they were built not just with language, but with *form-conscious intelligence*. They were designing for memory, not just meaning.

To treat symbols as time capsules is therefore to recognise that cultural transmission is not a by-product of civilisation—it is its *core technology*. The persistence of symbols like the mandala, the cross, or the serpent in global traditions points to a shared strategy for encoding durable knowledge. These symbols are not containers of static truth, but *platforms for cognitive reconstruction*. They compress complex insights into perceptual seeds, each capable of regrowth in new soil. Like ancient spores, they wait—not for decoding, but for engagement. And when the right mind meets the right symbol, meaning blooms anew, not as recovered fact, but as reactivated understanding.

The Grammar of Lost Languages

Language is not merely a tool for communication; it is architecture for thought. It constructs the mental scaffolding upon which we interpret reality, structure identity, and frame the sacred. When a language is lost, something more than syntax disappears—its epistemology vanishes with it. This is why so-called "lost languages," particularly those embedded in myth and mysticism, continue to provoke fascination not only among linguists and historians but among cognitive scientists, philosophers, and visionary seekers. They invite a deeper question: *What kind of mind creates a language not to describe the world, but to alter one's experience of it?*

Among the most provocative of these is the so-called Enochian language, recorded by the Elizabethan magician and mathematician John Dee and his associate Edward Kelley in the late 16th century. Claimed to have been transmitted by "angelic beings" during scrying sessions, Enochian presents a fully articulated grammar, syntax, and vocabulary. But its significance does not rest on whether it is an authentic relic of pre-Adamic speech—as Dee believed—but rather on *why* the language was constructed at all. It is an example of a *neuro-poetic system*—a symbolic architecture engineered to induce specific mental states, modulate perception, and encode sacred cosmology within phonetic structure. Enochian is not just a hypothetical spoken tongue. It is a *cognitive device*, a medium of transformation.

This raises a broader hypothesis: that certain "mythical" languages may have less to do with communication in the everyday sense, and more to do with *neural modulation.* Their structure, sound patterns, and symbolic density suggest they are designed to operate on the psyche—to trigger shifts in awareness, unlock archetypal content, or simulate states of ecstasy, communion, or otherness. These languages, then, are not dead; they are dormant systems—like software waiting for compatible hardware. They survive in manuscripts, inscriptions, chants, and visionaries' reports not because they were linguistically useful, but because they *reliably affected consciousness.*

We see similar characteristics in other obscure or allegedly "received" languages: glossolalia (or "speaking in tongues") among

Pentecostal Christians, the dense oral codices of Siberian shamans, and the incantatory speech of Vedic mantras. These forms are not designed for mutual comprehension in a transactional sense. Rather, they serve ritual, initiatory, or ecstatic functions. In glossolalia, the speaker is often in a trance state, and the vocalisation is seen not as gibberish but as an embodied utterance—one that bypasses rational syntax to access a more primal layer of expression. This idea has led some neurologists to propose that such speech phenomena are tied to unusual activation patterns in the temporal lobes and limbic system—regions implicated in language, emotion, and spiritual experience.

This neurological lens helps clarify why lost or "angelic" languages recur in visionary traditions. They do not emerge from a desire to catalogue the world. They emerge from altered states—dreams, trances, revelations—where conventional language fails. When Joan of Arc reported her divine voices, she claimed they spoke in a dialect she could understand intuitively but could not explain to others. Likewise, the Hopi elders maintain that their sacred chants are "not from this world," passed down in forms resistant to translation. In each case, the language is less about encoding information and more about modulating *attunement*—to divine presence, ancestral memory, or a sacred cosmological order.

This is not mere speculation. Recent studies in the cognitive sciences suggest that languages with complex phonotactics and recursive grammar—especially those involving syntactic inversion or semantic ambiguity—can stimulate broader neural integration than simpler, transactional languages. In this light, mythical tongues may function as *semantic psychedelics*—not by conveying new facts, but by reconfiguring the cognitive frameworks through which facts are understood. The grammar itself becomes a philosophical agent. In Enochian, for instance, word roots are deliberately arcane, and syntax is nested in non-intuitive structures, creating a sense of *defamiliarisation*. This pushes the practitioner into a heightened attentional state—exactly what one would expect if the goal were mystical realignment rather than communication.

There is also a structural feature worth noting: many mythical languages exhibit an unusual density of *symbolic redundancy*. That is, single phonemes or root words often carry multiple meanings across semantic domains—numerical, moral, cosmological, biological. This redundancy creates what cyberneticists call *error-tolerant systems*. If a word can mean "breath," "spirit," and "life-force" depending on intonation and context, then the language can survive mutation without losing core meaning. These are languages built to persist under stress—exactly what one would expect if their function were ritual preservation or transmission through unstable oral lineages.

We must also acknowledge the philosophical implications. If languages like Enochian or the so-called "angelic tongues" are mental architectures, then their existence blurs the boundary between invention and reception. Are they truly "received," or are they emergent properties of the human mind in altered states? Jung might call them *products of the collective unconscious*, while a modern neuroscientist might interpret them as the output of hyper-connected default mode networks under specific neuromodulation. But the convergence is noteworthy: both perspectives recognise these languages as *autonomous intelligences* within the psyche—syntactic entities with structure, purpose, and emergent logic.

From this vantage point, the obsession with "lost" or mythical languages becomes understandable. It is not nostalgia for forgotten grammars, but a deep intuition that *some languages encode more than speech*. They encode modes of being. This intuition drives everything from Tolkien's Elvish to the constructed languages of science fiction authors, some of which display internal consistency rivaling natural languages. These are not frivolous inventions—they are *synthetic mythologies*, where phoneme and cosmology are fused. The grammar becomes theology. The morphology becomes metaphysics. The reader or speaker does not merely decode; they undergo a form of initiation.

Perhaps this explains why language creation is often ritualised in myth. The gods speak first. The sacred language is always the *original* language. In the Book of Genesis, Adam names the animals—suggesting that language creates categories, and by extension, reality itself. In the Babylonian myth of Enuma Elish, language is weaponised—gods create and destroy with words. In Kabbalah, the Hebrew letters are not just phonetic units; they are *cosmic operators*.

This is not rhetorical flourish. It is a recognition that language is not simply descriptive—it is generative. It produces worlds.

The enduring allure of mythical tongues, then, is not their linguistic coherence but their *philosophical audacity*. They assert that speech can transcend communication and become communion. They are linguistic thresholds between the inner and outer cosmos, vehicles of metamorphosis rather than medium of consensus. They resist translation not because they are incoherent, but because their function is not semantic, but *transformative*. They are not meant to be understood—they are meant to *change the one who speaks them*.

Such languages, whether remembered, constructed, or transmitted in trance, are not failures of linguistic rationality. They are *extensions of consciousness*, probing the edges of what language itself can do. The real question they pose is not whether they were "real," but whether we are still capable of creating languages that *do more than describe*. Are we still willing to treat language as a sacred tool—a system for mapping not just what we know, but what we are becoming? If mythical tongues are time-resistant code, it is not because they survived history. It is because they point toward futures in which language is once again an *act of emergence*, not mere reflection. They are not relics. They are invitations.

Chapter 9
The Body Remembers — Myth in the Flesh

Much of contemporary myth analysis carries a quiet, almost unquestioned premise: myths are spun by the mind for the mind. Yet what if certain myths first arrive in the body? Not as polished stories, but as raw sensations—grief tightening the chest, a shiver coursing through the hands, a knot closing the throat, or that peculiar exhilaration felt at the knife-edge of danger. For all our symbolic parsing and taxonomic zeal, we may have misplaced a vital truth: myth was never merely a literary construct. It was lived. It was felt. It took root in the body before it ever reached the tongue. And if this is so, any genuine understanding of myth must begin not in the library, but in the nervous system.

Here, we take up the possibility that mythology is as much a somatic inheritance as it is a cultural or cognitive artefact—a system that encodes and transmits embodied knowledge across generations, shaped by both lineage and trauma. From the shuddering rapture of a shaman's initiation to the choreographed strikes of a martial kata, from sacred dances to the deliberate mark-making of ritual scars, the human body has always been myth's first canvas. Often, it is not merely the instrument that enacts the myth; it is the very soil from which the myth germinates, storing pain, memory, and revelation in flesh and fascia alike.

The sciences—neuroscience, epigenetics, somatic psychology—are slowly catching up to what ritual specialists have long known: the body can remember far more than the conscious mind allows. In trauma theory, Bessel van der Kolk has illuminated how overwhelming experiences that escape integration into narrative memory often reappear in purely physical form—muscle tension, sensory flashbacks, involuntary spasms. Trauma, in this view, is not just a mental wound but a disruption in the body's organising principles. This makes a startling parallel with ancient myths centred on bodily rupture, possession, and ecstatic metamorphosis. The scar of a warrior, the convulsions of a prophet, the god who is dismembered and remade—these are not idle embellishments. They are somatic imprints, retold in story precisely because they were first inscribed in the body.

Such correspondences prompt an uneasy question: might some myths be the direct offspring of generational trauma? Myth often gathers around disasters—deluges, exiles, massacres, pestilence. These need not be only cultural recollections. They may be biological legacies. Epigenetics now shows that trauma alters DNA methylation patterns, changes which can persist for multiple generations. A people who have endured famine, slavery, or siege may pass on not just cautionary tales, but physiological templates for vigilance, withdrawal, or suspicion. Myths that warn against hubris, exalt obedience, or vilify difference might, in this light, be less moral philosophy and more evolutionary survival code, sealed into story by the wax of memory.

Take the "wounded healer," an archetype scattered across cultures—from Greece's Chiron, who carries an incurable wound and becomes tutor to heroes, to the Zulu isangoma, whose calling often comes through sickness or psychic fracture, resolved only by initiation. Neuroscience would recognise here the principle of neuroplasticity in crisis: the breaking apart of established pathways to allow for profound reorganisation. In such myths, healing springs not despite the wound but through it. The body is torn open, and new circuitry grows. Myth here is not a decorative metaphor—it is the map of the process itself.

Ritual disciplines make this interface almost impossible to ignore. Yogic breath control, monastic fasting, West African trance dances—each deliberately alters the nervous system, shifting vagal tone, changing brainwave patterns, triggering cascades of hormones. These are not incidental by-products; they are the generative engine of mythic vision. Whether in the vision quest, the sweat lodge, or the whirling of dervishes, the body enters an altered state and begins to perceive in patterns—myth unfolding not as invention, but as the recognition of a deep structural rhythm linking inner and outer worlds.

Modern neuroimaging now affirms what initiates have embodied for centuries. Long-term contemplative practice thickens cortical layers, heightens interoceptive awareness, and strengthens the brain's capacity to regulate fear. But the shaman who has fasted, fought, danced, and bled would have little patience for this "discovery." They have always known that story, vision, and restoration arise not from intellect alone, but from a patterned choreography of the flesh. The body, far from impeding transcendence, is its doorway.

This suggests a fresh approach to myth: a somatic hermeneutics. Instead of only parsing symbols or tracing historical analogues, one might ask: what physical state does this myth evoke or describe? What sensation is conjured in its telling? What rhythm or breath underlies its narrative pulse? Seen this way, even the most abstract myths reveal themselves as sequences of bodily experience, choreographed into narrative but rooted in sensation.

Consider how dance encodes myth in muscle and bone. In Balinese barong performance, the eternal play of chaos and order is not merely recited—it is enacted through ritualised movement refined over generations. The gestures themselves are a form of knowledge. In Brazilian capoeira, every sweep and feint carries the residue of resistance, survival, and cultural fusion, transmuted into a playful but deadly art. Such traditions are more than spectacle; they are archives of lived history, stored in tendons and breath.

This reframing extends to the way we read illness, injury, or disability in myth. The limp of a hero, the delirium of a seer, the scars of a shaman—these need not be metaphors for inner conflict, but acknowledgements of initiation through difference. Rather than pathologising the atypical body, myth often consecrates it. It reminds us that rupture and repair are not markers of failure but stages of transformation. To heal is not to return to what was, but to become something new, more attuned.

"Myth in the flesh" is no poetic abstraction. It is what happens when grief buckles the knees, when dance overtakes the thinking mind, when a tale pierces the gut before the brain can name it. It insists that thought does not reside solely in the head, that memory is not bound to chronology, and that healing can occur without a single word. Dreams, visions, and metaphors may ignite myths, but they are proved in the tissues. The breath shallows before the danger is named; the stomach knots before the memory surfaces. The body knows. Myth may be, at its root, the body speaking in its own tongue.

What follows is an attempt to trace this subterranean language—how sensation shapes myth's structure, how inherited trauma leaves both genetic and symbolic traces, how movement inscribes story in muscle, and how ancient "superstitions" find vindication in modern labs. This is no regression to the primitive, but a recalibration: an insistence that the intellect is not enough. Myth was never solely cerebral. It is a teaching embedded in breath and scar, in gait and chant. To engage it fully, one must feel it—not merely as metaphor, but as method; not only as legend, but as living legacy.

"Somatic Mythology"

Mythology has traditionally been treated as literature, its symbols decoded through textual exegesis, linguistic parallels, and archetypal theory. But this approach rests on a silent assumption—that the body is secondary to the intellect, that myths are best understood with the mind alone. This is a limitation not shared by the traditions that birthed these myths. In cultures from West Africa to Polynesia to the Arctic, myths were not just *told.* They were *danced, sung, fought, trembled, and endured.* They lived in muscle memory, vocal resonance, ritual posture. Before myth was etched in parchment or chiselled in stone, it was carried in the chest and limbs. The myth was not a metaphor. It was a movement. And it moved through the body first.

To speak of "somatic mythology" is to radically re-centre myth in the sensorium: to recognise that stories about gods, monsters, and heroes often mirror not just social or cosmic structures, but *bodily experience.* This shift changes how we interpret fear, awe, transformation, and ritual. It allows us to ask not only what a myth *means*, but what it *feels like.* And it challenges the assumption that symbolic content is the only carrier of truth. The human body, with its tension and release, scars and scars remembered, may be both the instrument and archive of mythology.

Consider fear—not the abstract concept, but the visceral contraction it creates in the gut, the prickling skin, the breath held too long. These sensations are consistent across cultures and time periods. They are foundational to the human experience. Is it any surprise, then, that mythic structures are often built around terror? Dragons, underworlds, night-journeys, labyrinths—these are not just plot devices. They externalise *internal physiological states.* The slaying of the beast is the pacification of the sympathetic nervous system. The descent and return mirror the cycling of parasympathetic restoration. A myth is not just a way of making sense of the world—it is a way of metabolising what the body cannot bear alone.

Many ritual traditions make this explicit. In Balinese *topeng* theatre, performers don sacred masks to enact mythological cycles. But to wear the mask is not merely to act. It is to *become inhabited.* The performer's breath changes. The spine aligns differently. The mask reshapes perception, physically and neurologically. Similarly, in Japanese Noh theatre, the controlled, glacial movements of the actors are not for aesthetic pleasure alone. They induce *altered states* in both performer and audience—attuning nervous systems through rhythm, gravity, and micro-movement. Myth is not being performed. It is being *channelled* through somatic architecture.

Shamanic traditions worldwide reveal even deeper links between body and myth. The initiation of a Siberian shaman often includes dismemberment in vision—bones stripped, organs rearranged, the body reconstructed with new materials. Anthropologists have long debated whether these images are metaphor or hallucination. But this misses the point. The visionary event corresponds with real, felt changes in bodily awareness. The shaman emerges changed not just ideologically, but physiologically. Brainwave patterns shift. Hormonal cascades follow. In South American ayahuasca ceremonies, the *icaros*—songs sung by the shaman—are not entertainment. They are *vibrational tools*, designed to direct the visionary state by modulating the auditory and vestibular systems. Myth is being constructed in real time, through the body's sensory channels.

Even martial arts, often seen as systems of combat, function in many cultures as *mythic expression.* The movements of Chinese Tai Chi

Chuan, for instance, are not simply defensive postures. They are cosmological metaphors in motion: Yin and Yang spirals enacted through breath and bone. Capoeira, developed by enslaved Africans in Brazil, encoded resistance into fluid motion. It is both story and body—trickster myth enacted through embodied feint and flow. In these systems, the hero's struggle is not narrated—it is *practised*, *rehearsed*, and *remembered* in the musculature.

Such examples illustrate that ritual movement often carries cognitive and emotional information that precedes language. This aligns with discoveries in contemporary neuroscience, which increasingly acknowledge the primacy of the *embodied brain*. Theories of *embodied cognition* argue that the brain evolved not merely to process symbols, but to coordinate action in space. Meaning, in this framework, is not abstract but *enacted*. Stories don't simply reside in cortex—they are shaped by posture, rhythm, proprioception. Myth, then, may be the cultural crystallisation of *action-patterns* that proved adaptive—narrativised to preserve what could not yet be spoken.

Somatic markers theory, developed by neurologist Antonio Damasio, supports this view. Emotions arise from the body's evaluation of internal and external stimuli, and these emotional responses—marked in the viscera, heartbeat, muscle tone—are what guide decision-making. In myth, characters are often faced with decisions of moral or existential weight. The body responds to these dilemmas *before* the intellect. The hairs on the neck rise before the warning is heeded. The stomach tightens before the betrayal is revealed. These somatic cues are not narrative flourishes. They are *core to the myth's logic*. The reader or listener experiences the story not just cognitively, but *biophysically*.

This embodied resonance also explains why certain stories persist. They are not remembered because of literary quality, but because they trigger deeply ingrained somatic responses. Consider the motif of descent and return—from Inanna's underworld journey to Christ's harrowing of hell. These stories map closely onto patterns of *trauma and recovery*. The nervous system experiences overwhelm, collapses into immobility, and—if it can—reorganises and re-emerges. Trauma researcher Peter Levine's work on somatic experiencing is instructive here: recovery is not about remembering the event intellectually, but *discharging* its trapped energy through sensation and micro-movement. Myth does something similar. It enacts, in symbolic form, the body's path through disruption, fragmentation, and reintegration.

This view demands we reinterpret what a "ritual" is. A ritual is not performance art. It is a structured environment in which the body undergoes *controlled stress*—exertion, breath control, social vulnerability—in order to induce recalibration. Whether it is the fire-walking of Polynesian initiations, the trembling convulsions of Pentecostal worship, or the breathless stillness of Zen meditation, rituals work because they *alter physiology*. And in doing so, they make possible new mythic insight. The hero's transformation is not symbolic—it is a *neurophysiological reality*.

Moreover, somatic mythology reframes how we understand pathology. Many individuals experience intense bodily sensations—panic attacks, dissociation, spontaneous movements—that do not conform to Western psychiatric models. These may be pathologised as disorders. But traditional cultures often interpreted them as *mythic callings*. The shaking novice was not labelled epileptic or hysteric. They were recognised as touched by the divine, or summoned by ancestors. The body's disruption was not evidence of illness, but *invitation*. This is not to romanticise suffering, but to ask whether certain bodily phenomena have been misread in clinical contexts precisely because we have lost the mythic lens.

This also has political implications. If myth lives in the body, then access to mythic understanding is *not* confined to the literate or scholarly. It belongs to dancers, fighters, mothers, the grieving, the ecstatic, the displaced. It is democratised through sensation. In this view, marginalised groups often hold vital mythic knowledge—not because of their oppression, but because of their *somatic insight*. Their bodies have moved through realities that others have only read about. And in their motion, myth has been retained.

Somatic mythology therefore calls for a new methodology in both scholarship and personal practice. It asks us to track myth through breath and gesture, to feel which stories expand or contract the chest, which movements invite trance or terror. It challenges scholars to *feel* their theories, to let the text speak through muscle and bone. And it invites every reader to ask not just "What does this myth mean?" but also "Where do I feel it?"

The myths that endure are those that resonate in the body. Not because they were well-written, but because they were *well-lived*—tested and tempered in breath, in blood, in tension and release. If myth is a map, the body is the terrain it charts. If myth is a code, the body is the key. And if myth is a path to transformation, then the body is not the obstacle—it is the guide. Understanding this is not merely interpretive. It is *participatory*. To truly grasp myth, one must move with it. And let it move through.

"Trauma Carried Through Generations":

When a child of genocide survivors dreams of fire, fleeing, or faceless enemies—despite never having seen war—what exactly is being remembered? When mythologies across disparate cultures begin with floods, with sons killing fathers, with the chaos that preceded order, are these simply archetypes—or could they be *echoes* of real, embodied pain, transmitted across generations not just by word or symbol, but by biology itself? The field of epigenetics is reshaping our understanding of memory, inheritance, and myth, suggesting that trauma doesn't simply disappear with the generation that experiences it. It embeds, encodes, and adapts—leaving traces in behaviour, physiology, and narrative structure. Myths, under this lens, may be more than cultural artefacts: they may be *survival blueprints*, forged in the heat of ancestral fire, carried forward in both blood and story.

Epigenetics refers to biochemical mechanisms—such as DNA methylation or histone modification—that regulate gene expression without altering the DNA sequence itself. These processes are responsive to environmental conditions, including extreme stress, malnutrition, violence, and disruption. Importantly, certain epigenetic changes have been observed to persist across generations. Studies on the descendants of Holocaust survivors, for instance, have shown altered cortisol levels and stress responses, even when those descendants were raised in stable, non-traumatic environments. Similar patterns have been found among grandchildren of those who lived through famine or forced displacement. This is not a metaphor. It is a measurable, physiological inheritance of history.

But what happens when science meets story? Myths—especially origin myths—often carry signatures of catastrophe: deluge myths after global floods, tales of fire and exile following volcanic eruptions, or narratives of divine punishment following social collapse. It is tempting to view these stories as moral allegories. But perhaps they are also *epigenetic codices*, mapping survival strategies and psychological responses that were too overwhelming to be stored in conscious memory alone. The catastrophic event may end, but its imprint continues—stored in stress systems, passed through hormonal pathways, and given narrative form in the cultural imagination.

One of the most studied examples is the widespread myth of the Great Flood, appearing in Mesopotamian, Hindu, Aboriginal, and Native American traditions. These stories often include divine judgement, mass death, and a chosen survivor—Noah, Manu, Deucalion—who carries forward the seed of renewal. While climate data supports the existence of massive post-glacial sea level rise around 12,000 years ago, what is striking is not just the myth's recurrence, but its psychological tone: terror, helplessness, grief, and rebirth. These are the emotional markers of trauma. The flood myth may be a kind of epigenetic memory theatre—a way of encoding the unprocessable through story, so that it can be carried, rehearsed, and perhaps, in time, transformed.

This also helps explain why certain mythic patterns seem so disproportionate or grotesque. Why do so many myths include themes of incest, dismemberment, or cosmic cannibalism? From the Greek story of Cronos devouring his children to the Aztec goddess Coatlicue birthing both life and death, these narratives may reflect distorted echoes of real generational wounds—child sacrifice, starvation, societal collapse—abstracted into symbolic forms that allow communities to face what would otherwise be psychically unbearable. Myth, in this view, is not just imagination. It is *somatic intelligence rendered symbolic*—a bridge between the biology of survival and the culture of continuity.

Modern psychology offers further support. Research in trauma therapy, particularly among clinicians like Bessel van der Kolk and Gabor Maté, has revealed that trauma is stored not primarily in narrative memory, but in the body—in the autonomic nervous system, in behavioural patterns, in stress reactivity. When unresolved, trauma tends to replicate itself across generations—not by storytelling, but by *unconscious re-enactment.* Families of trauma survivors often display patterns of hypervigilance, emotional suppression, or overcompensation, even when the original context is long gone. Myths can be seen as a cultural analogue to these patterns: *symbolic re-enactments* of unresolved collective trauma, playing out in ceremonial form to allow some measure of catharsis or re-regulation.

Ritual plays a key role here. In many traditional cultures, myths are not simply told—they are *enacted* through dance, fasting, song, and ordeal. These embodied practices serve to *discharge* inherited trauma, providing structured ways to revisit and transform ancestral pain. The Dagara people of West Africa, for instance, use grief rituals to process not just personal sorrow but intergenerational wounds. The Maori haka, far from being merely a war dance, is also a somatic invocation of lineage, defiance, and resilience. These are not theatrical performances. They are neurological interventions—strategies for metabolising inherited stress through collective rhythm, breath, and movement.

It's crucial to avoid oversimplification. Not all myths stem from trauma, and not all trauma becomes myth. But the overlap is compelling. In psychoanalytic terms, Carl Jung's concept of the *collective unconscious* suggests that myths represent shared psychic structures shaped by the deep evolutionary past. Today, we might translate this idea into *epigenetically-informed neurobiology*—where archetypes are not mere symbolic forms, but emergent properties of nervous systems shaped by common environmental pressures over generations. The hero's journey, the underworld descent, the confrontation with the shadow—these may all represent *neurally embedded strategies* for navigating survival challenges, honed over millennia, stored in genes and refined in story.

Importantly, this framework doesn't reduce myth to mere pathology. It elevates it. To recognise myth as a carrier of trauma is not to diminish its beauty or power, but to understand *why* it matters—why it grips us, why it survives, why it returns. Myths that endure do so not because they are aesthetically pleasing, but because they *speak to structures already present in us.* They align with the stress architecture of our ancestors. They echo the hormonal tides of loss and resilience. And in doing so, they offer both diagnosis and possibility: a map not just of what happened, but of how to move forward.

This has implications for fields far beyond mythology. In psychotherapy, recognising the mythic dimension of trauma can enrich both diagnosis and healing. In education, teaching mythology as a *living system of inherited wisdom*—rather than distant fiction—can restore its relevance. In public health, understanding intergenerational trauma through both scientific and mythic lenses may provide more culturally attuned interventions. Even in policy, acknowledging the lingering epigenetic effects of colonialism, slavery, and forced migration reframes debates about responsibility, reparations, and justice.

Ultimately, myths that carry trauma are not tragedies to be dissected. They are acts of resilience. They are data sets encoded in metaphor, capable of being retrieved, re-read, and rewritten. They show us that pain need not vanish to lose its power. It can be transformed. It can become wisdom. But only if we are willing to feel its weight, trace its lineage, and listen not just to the words—but to the *bodily frequencies beneath them.*

To do that, we must move beyond the page. We must hear the tremble in the voice of the elder recounting a half-remembered tale. We must observe the tightening jaw, the unshed tears, the rituals repeated without knowing why. And we must approach myth not as a relic of superstition, but as a highly efficient system of embodied knowledge, carrying lessons not just from the ancestors' minds, but from their *nervous systems.* When we read myths in this way, we are not escaping history. We are entering it—fully, somatically, and with the radical humility that we, too, are still living out stories older than we know.

"The Ritual Wound":

Why do the greatest mythological heroes never emerge unscathed? From Achilles' vulnerable heel to Odin hanging himself from Yggdrasil to gain wisdom, the narrative of the heroic figure is not one of unbroken strength, but of pain endured, survived, and transmuted. The hero does not triumph by remaining intact; the hero transforms by being *broken open.* This recurring theme is not incidental—it reflects a deep neurological, psychological, and social truth: *growth requires rupture.* The mythic wound is not a flaw in the tale. It is the mechanism by which change becomes possible. And today, neuroscience is beginning to confirm what myth has always encoded—that the wound is not only where the light enters, but where the brain rewires, adapts, and evolves.

The wound, in myth, is not merely physical. It often represents a breach in identity, certainty, or control. When Prometheus is bound and tormented for giving fire to humanity, or when Osiris is dismembered and scattered across the land, the motif is not about simple punishment—it is about symbolic death and reintegration. These stories are not masochistic dramas. They are neurological blueprints. Neuroplasticity—the brain's capacity to rewire itself in response to experience—depends on *disruption.* The nervous system does not transform during equilibrium. It transforms when existing models break down. In mythic terms: the hero is wounded so that a new pattern can be born.

Contemporary trauma research reinforces this logic. The process of healing from psychological wounds is not linear. It involves a shattering of assumptions, a confrontation with loss or vulnerability, and the emergence of new ways of seeing and being. Post-traumatic growth, a field of study in psychology, refers to the adaptive changes that can occur after adversity: increased resilience, spiritual depth, and a renewed sense of meaning. This mirrors what ancient stories already tell us—that the path to integration goes *through* fragmentation, not around it. The ritual wound, then, is not a tragic setback. It is the *catalyst.*

Consider the symbolic structure of the hero's journey as outlined by Joseph Campbell and later reframed by cognitive theorists. The hero leaves the known world, faces a crisis, is symbolically or literally dismembered, and returns changed. What often goes unremarked is that this structure maps not only mythic plots but neurological transformation. The act of crossing a threshold—into battle, into exile, into unknown experience—provokes stress, destabilisation, and vulnerability. It is *after* this liminal phase, when the nervous system has been flooded and reset, that integration becomes possible. The 'boon' the hero brings back is not just knowledge; it is *rewired consciousness.*

The ancient ritual traditions recognised this implicitly. In initiation ceremonies across cultures—from Aboriginal walkabouts to Maasai lion hunts to the ritual scarring of the Sepik River peoples—pain is deliberately invoked. Not for sadistic ends, but to mark the passage from one cognitive and social state to another. The wound in this context is *sacred.* It creates an opening—neurologically, emotionally, culturally—through which transformation flows. The initiate does not merely learn; they are changed. And this is precisely why rites of passage were never purely symbolic. They were *visceral,* because the brain learns what the body remembers.

This is equally evident in martial and meditative traditions. The repetitive, embodied strain of capoeira, kalaripayattu, or Shaolin kung fu is designed not only to train skill, but to reshape perception and identity. Pain is not an obstacle to mastery; it is the crucible. Similarly, in monastic settings, long periods of fasting, sleeplessness, or silence induce altered states of consciousness. The 'wound' here is metabolic or sensory, disrupting homeostasis to enable shifts in awareness. These ancient disciplines—long dismissed as superstitious or punitive—are now being re-examined through the lens of neuroplasticity and embodied cognition. What myth and ritual have long honoured, neuroscience is finally beginning to map: *transformative change is encoded in the body before it is narrated by the mind.*

This theme also explains the recurrence of the wounded healer archetype—figures like Chiron, who cannot heal his own wound but becomes a teacher to others. The scar becomes the site of wisdom. Carl Jung recognised this dynamic when he argued that the therapist must be wounded to understand the suffering of the patient. In a wider cultural sense, this mirrors how societies remember trauma—

through commemorations, physical monuments, and embodied rituals. The scar is kept visible not for spectacle, but for transmission. The community remembers *through* the wound.

But not all wounds are transformative. Some paralyse. Some entrench suffering. This is where myth provides an ethical guide. The hero who is wounded and *does not return* becomes a cautionary figure—like the mad prophet, the ghost, the exile. Their failure is not in being wounded, but in becoming identified with the wound itself. The mythic cycle insists that transformation must follow injury—that descent is only complete when it is followed by *return*. This speaks to a profound neurological and psychological truth: *plasticity without integration becomes disorganisation*. Change must be metabolised. Pain must be given context. And myth, at its best, provides that container.

Modern clinical practice is, in some ways, catching up to this ancient insight. Therapies such as somatic experiencing, EMDR (eye movement desensitisation and reprocessing), and narrative therapy work precisely by *re-engaging* with the wound—emotionally, bodily, narratively—without being overwhelmed by it. These approaches mirror the mythic structure: the descent, the confrontation, the meaning-making. The wound is not avoided. It is *revisited*, with new tools, new awareness, and new witnesses. Healing, then, is not the erasure of pain. It is the reintegration of pain into a coherent, adaptive identity.

It is important to note that not all cultures read the wound as suffering. In many Indigenous frameworks, the wound is a *gift*—the place where the human meets the divine. Vision quests, sweat lodges, and isolation fasts are structured to induce suffering not for punishment, but for revelation. In such contexts, pain is not pathology. It is pedagogy. The body, temporarily disoriented, becomes a conduit for new insight. And the story that emerges from that pain is not just personal. It is *ancestral*, *communal*, and often *cosmic*. This is the radical power of the ritual wound: it connects the personal nervous system to the mythic nervous system of the collective.

We must also consider the ethical dimension of this motif. If the wound is the site of transformation, what are the implications for systems—religious, medical, educational—that *inflict* wounds in the name of growth? Myth is often misused to justify unnecessary suffering. The difference lies in *intention* and *integration.* A wound inflicted without guidance, without container, without meaning, is violence. A wound engaged with in a ritual context, with support and integration, becomes medicine. Myth does not glorify pain. It *contextualises* it.

The contemporary fascination with "resilience" often strips away this nuance, treating pain as an individual failing or success. But myth reminds us that transformation is not solitary. The hero is wounded *within* a narrative. The transformation is *witnessed.* There is a community that receives the new self. Neural rewiring happens in *relational contexts.* Our brains change not in isolation, but through mirrored presence, shared story, and communal meaning. The ritual wound, then, is not just personal. It is social.

To read myths through this lens is to reclaim them not as fiction, but as functional models of change. The wound in myth is not a metaphor. It is a neuro-symbolic event, reflecting what we now know about learning, trauma, and plasticity. The stories endure because they are *accurate.* They map not only the outer world but the internal process by which humans adapt, repair, and evolve. And as modern societies re-examine the role of ritual, embodiment, and story in healing, the mythic wound becomes not a relic of superstition—but a tool, a guide, and a mirror.

Ultimately, the question is not whether we will be wounded, but *how we will relate to the wound.* Will we deny it, become it, or transform through it? Myth gives us a template. Neuroscience gives us a mechanism. The challenge is to bring the two into dialogue—not to romanticise suffering, but to understand it as part of the process by which *the nervous system writes new code.* The wound is not the end of the story. It is where the real story begins.

"Biological Echoes of the Sacred":

What modern neuroscience is beginning to uncover, ancient ritualists may have already known through direct experience.

Practices like fasting, rhythmic breathwork, and sensory isolation—hallmarks of religious and spiritual traditions across the globe—were not just symbolic acts of devotion. They were precise, embodied techniques for altering consciousness, activating neuroplastic processes, and provoking states of insight, surrender, and transformation. Long treated as mystical or purely religious, these states are increasingly being studied not as anomalies, but as *biological capacities*. The sacred, it appears, leaves measurable footprints in the body. And those footprints offer a map not just of spiritual ecstasy but of neurocognitive recalibration.

One of the most striking examples of this convergence between ancient practice and contemporary science is the growing body of research around breathwork. Techniques like *pranayama* in Indian yoga, *tummo* breathing among Tibetan monks, or the controlled hyperventilation used in modern Holotropic Breathwork were traditionally used to regulate vital force—*prana*, *chi*, or *ruach*. Today, we understand these not merely as metaphysical concepts but as functional descriptors of altered physiological states. Controlled breathing directly influences the autonomic nervous system, modulating heart rate variability, reducing cortisol, and even shifting brainwave activity from beta to alpha and theta states—associated with introspection, creativity, and expanded awareness.

The neurobiological basis of this is not speculative. Studies using functional MRI and EEG have shown that specific breathing rhythms activate the insula, a brain region responsible for interoception—our perception of internal bodily states. This supports the idea that ancient breathwork practices cultivated not only stillness or trance, but *embodied self-awareness*. Furthermore, sustained breathwork has been linked to increases in gamma coherence—neural synchrony associated with insight, unitive experiences, and 'flow states'. These findings align uncannily with anecdotal reports from practitioners who describe states of egolessness, visionary insight, or mystical union following intensive breathing rituals. The techniques that once prepared shamans to meet gods or ancestors may, in neurological terms, be precision tools for disrupting default-mode cognition and initiating neurophysiological reset.

Fasting—another core element in religious tradition—is equally rich in implications. Across cultures, fasting has been used not merely as an act of purification, but as a gateway to altered perception. From the forty-day fasts of Moses and Jesus to the Ramadan of Islam and the vision quests of Indigenous cultures, voluntary abstinence from food was a mechanism of transcendence. Today, researchers are uncovering its biochemical mechanisms. Prolonged fasting induces *ketosis*, a metabolic shift that affects not just energy systems but neurochemistry. Ketones like β-hydroxybutyrate influence brain-derived neurotrophic factor (BDNF), a protein critical to neuroplasticity, learning, and memory.

More importantly, fasting initiates a complex adaptive response in the brain—heightened alertness, increased neurogenesis, and stress resistance. Evolutionarily, this made sense: a hungry animal must think sharply and move efficiently to survive. But religious traditions appear to have *ritualised* this biological response into a tool for spiritual refinement. The lucidity, emotional clarity, and visionary states often reported during fasting may stem not from divine intervention, but from neurochemical optimisation. In this light, the ascetic is not just a moral example, but a biohacker of antiquity—tuning bodily systems to access otherwise latent potentials of mind.

Altered states induced through sensory deprivation, chanting, and trance dancing add another layer. When Sufi mystics spin in whirling meditation, when Amazonian shamans chant for hours under the influence of ayahuasca, or when indigenous Siberian reindeer herders engage in ecstatic drumming, they are engaging in *entrainment*—synchronising internal rhythms with external stimuli to shift consciousness. Contemporary neuroscience recognises this phenomenon in terms of *brainwave entrainment*—a process by which external rhythmic input can influence the frequency of brain activity.

These states often involve a breakdown in ordinary sense of time, self, and space—experiences traditionally interpreted as divine contact or spiritual revelation. Yet fMRI and EEG studies show that such states involve decreased activity in the *default mode network*—the brain system responsible for narrative self-reference—and increased connectivity between sensory, emotional, and memory-processing areas. The result is often described as a feeling of vastness, presence, or being "one with all things". Ancient rites did not just mimic transcendence. They *produced* it—reliably, physiologically, and often collectively. And what modern scientists call *neural integration*, ancient priests might have named as *grace* or *revelation.*

This convergence challenges the historical dichotomy between the sacred and the scientific. It suggests that ancient ritual was not merely a response to the unknown, but a deliberate *engagement* with the neurobiological architecture of human transformation. And it reframes the sacred not as an abstract ideal, but as a *state-dependent reality*, accessed through reproducible techniques and embodied disciplines. To feel awe, to dissolve into unity, to perceive a world more vast and interconnected than the ordinary mind can grasp—these are not illusions. They are *capacities*, encoded in the human nervous system and accessed through patterned disruptions.

That these disruptions were almost always ritually structured is no coincidence. Ritual provides the context and containment necessary for navigating altered states without disintegration. Breathwork without guidance can induce panic. Fasting without intention can become disorder. The ancient traditions understood that transformation required *safety and structure.* The priest, the shaman, the elder served not only as initiator but as *regulator*—helping the participant interpret, metabolise, and integrate what they encountered. In psychological terms, this is *scaffolding.* In neurological terms, it is guided plasticity. The sacred, then, is not simply what overwhelms; it is what *reorganises*—under the right conditions.

The contemporary resurgence of interest in psychedelics is a relevant parallel. Compounds like psilocybin and DMT are now being studied for their therapeutic potential, particularly in cases of trauma, depression, and existential distress. What these substances do—biochemically—is destabilise existing neural hierarchies, allowing suppressed memories, emotions, and insights to surface. But what is equally important is the *context.* Clinical trials pair psychedelic exposure with *preparation, intention, and integration*—exactly the scaffolding provided in traditional ritual. The medicine is not the molecule alone. It is the *ritualised encounter.*

The implication is profound. What if much of what we call religion is a *coded protocol* for invoking and integrating altered neurobiological states? What if the sacred is not a metaphysical domain but a *reliable consequence* of engaging the nervous system in specific patterned ways? This would explain the universality of such practices—why fasting, chanting, breathwork, and solitude appear in cultures otherwise separated by oceans and centuries. They were not copying each other. They were discovering what the body already knew: that transformation is not a matter of belief, but of *physiology*.

And what if the symbols, myths, and cosmologies built around these practices were not distractions, but *maps*—cognitive frameworks for navigating and integrating the non-ordinary insights provoked by altered states? The hero's descent, the prophet's vision, the martyr's ordeal—these are not just stories. They are *sequences*. They mirror the phases of neuropsychological transformation: destabilisation, revelation, reintegration. Myths persist not because they are comforting, but because they are *biologically plausible*. They match what it feels like to be torn open and rewoven.

This raises an important ethical question: how do we reclaim these biological echoes of the sacred in modern life without reducing them to tools of optimisation or performance? The ancient practices were not simply techniques. They were embedded in meaning, community, and cosmology. Stripped of that, altered states risk becoming disorienting or addictive. The science confirms the power of these practices. But *wisdom lies in their integration*—in remembering that breath is not just oxygen exchange, but orientation. That fasting is not deprivation, but dialogue with desire. That silence is not absence, but a space where awareness becomes audible.

In this synthesis of science and spirituality, we are not inventing something new. We are *remembering* something old—something encoded in the rituals of our ancestors, and now legible in the data of our instruments. The body, when treated with rhythm, intention, and reverence, becomes a site of revelation. The sacred is not elsewhere. It is *endogenous*. And the echo is not fading. It is returning—with clarity, with evidence, and with the possibility of a new kind of literacy: one that reads the flesh as scripture, and the nervous system as myth in motion.

Chapter 10
Mirrors to the Future — Myth as Prophecy

We often consign myth to the dust of antiquity—an heirloom from an era before logic, before data sets and predictive models. Yet, the more closely one studies it, the more a provocative thought begins to take shape: perhaps myth has never been solely concerned with our origins, but with our trajectory. Not prophecy in the crude sense, but a disciplined sensitivity to the shapes and rhythms of change—a kind of cultural foresight that pre-dated our formalised methods of scenario planning or risk modelling. In that light, myth might be seen as humanity's first technology of anticipation: a vessel for warnings, adaptive metaphors, and simulated catastrophes long before the language of algorithms and probabilities existed. Maybe the ancients did not so much *foretell* the future as *recognise* it when it appeared.

Across continents and centuries, certain images and plotlines return with unnerving regularity—floods swallowing entire realms, creations rebelling against their makers, pestilences scouring the land, divine wrath taking forms not unlike environmental collapse. Such motifs are not scattered accidents; they appear with such frequency and persistence that they suggest something closer to cultural memory—an archive of premonition embedded in narrative form. The Mesopotamian tale of Atrahasis and the Biblical Flood each describe humanity's overabundance prompting a divine "reset." The Mahabharata's weapons bear an uncanny resemblance to nuclear annihilation. In Chinese Daoist thought, one finds machines that think; in Jewish folklore, golems rise; in alchemical writings, homunculi stir to life. All seem strangely in step with crises that now accelerate under the flag of modernity.

What is striking is not that these old tales are "accurate" in detail, but that they mirror structural truths. Ancient storytellers did not need meteorology to notice the cycle of plenty, imbalance, collapse, and renewal. Instead of statistics, they left us metaphors—fire and flood, barren soil, skies that refuse rain, gods in retreat—that capture the essence of systemic fragility. Their intimacy with soil and season, their reliance on cyclical abundance, honed a pattern-sense that was both ecological and existential. That we now find those same patterns in the lexicon of climate science is less mysticism than evidence of deep, generationally tuned perception.

Artificial intelligence, too, has its mythic foreshadowing. From the bronze giant Talos to Pygmalion's breathing statue, from the clockwork marvels of the *Book of Ingenious Devices* to the eerily autonomous beings of medieval imagination—there runs a thread of fascination and dread. The detail is less important than the central question: what does it mean to create a mind that does not dream our dreams? Unlike the plough or the steam engine, such an invention crosses into the terrain of the uncanny. The Other is no longer distant; it is forged by our own hands. Long before machine learning, myth understood the ethical vertigo this would provoke. Today's debates about bias, agency, alignment, and runaway intelligence are simply modern inflections of an older anxiety.

Apocalypse, too, is worth revisiting. We tend to see it as primitive fear—the fire, the tempest, the fall. But many apocalyptic traditions are subtler than that. Ragnarök is not final obliteration but a prelude to renewal. The Hindu Yugas cycle from decline to rebirth. The Aztec cosmology tells of successive suns, each undone by a different element, only for another to rise. These are not merely tales of ending; they are meditations on entropy and return. Such stories give cultures a way to contemplate systemic collapse without surrendering to despair—a mental architecture for living through turbulence. In an era of planetary strain and institutional unravelling, that architecture may be as necessary as any piece of hard science.

Systems theorists like Norbert Wiener and Donella Meadows have argued that complexity has predictable fault lines—feedback loops, tipping points, phase changes. Myth, in its narrative way, charted similar territory. Icarus's fall, Babel's crumbling tower, Eden's exile—these are not quaint moral lessons, but durable maps of overreach and fracture. Their survival is no accident. They track the thresholds of sustainability, the brittleness of human ambition, the strange reflex to grasp for the godlike and stumble. If myth has been dismissed as primitive fiction, perhaps it is our interpretive frameworks, not the myths themselves, that have been impoverished.

The prophetic force of myth lies not in pinpoint forecasts, but in its grasp of consequence. These tales are feedback systems rendered in story form: if you draw too heavily on what sustains you, if you ignore the limits that bind you, something—soil, spirit, or system—will fail. They feel like memories even when we know they are not factual because they encode patterns we've carried for centuries. They are, in a sense, memory devices—simulations of collapse and recovery woven into narrative fibre.

Some contemporary scholars now propose that myth be understood as a form of strategic foresight—not the sterile charting of probable futures, but an embodied attunement to emergent patterns. In this frame, the mythmaker is less entertainer and more cultural seismograph—sensing tremors in the social, ecological, and spiritual strata. Myth lacks data points, yet compensates with breadth: it folds politics into ecology, psychology into cosmology. And by working through symbol rather than argument, it slips past the defences of analysis into intuition, ritual, and feeling. It does not predict; it prepares.

If we approach myth in this way, it becomes more than a literary curiosity. It could function as a survival tool. In times when linear growth models falter and the predictable collapses into chaos, narrative—especially one that can hold uncertainty without resolution—may become as vital as measurement. Myths, far from relics, are cognitive instruments, tuning us to resilience and guiding us through unmapped terrain.

In the end, the question is not whether ancient peoples could somehow "see" us coming, but whether we can still see ourselves with their clarity. Myth was once a mirror; now, it might serve as a weather vane. The choice of what to do with it—whether to file it under entertainment or treat it as a kind of operating manual—belongs to us. For myth, at its heart, is about what endures: the patterns that press and repeat, the warnings that return until we choose to heed them. That persistence is not accident; it is design. And it is, perhaps, the oldest kind of foresight we possess.

Did the Ancients Predict Us?

We often assume that our age is radically unlike any that came before it—defined by machine intelligence, synthetic biology, global climate change, and the looming possibility of planetary-scale transformation. Yet beneath the surface novelty of our technological moment lies an uncomfortable question: *how new is all of this, really?* Could it be that some of the stories we dismiss as ancient superstition were, in fact, early pattern-recognition systems—primitive yet profound attempts to anticipate the recursive challenges of civilisation itself? If so, then the ancients may not have *predicted* us in the magical sense. They may have *modelled* us—in narrative code.

This chapter does not ask whether the myths of old contained supernatural foresight. It asks something more unsettling: did they encode long-term feedback loops that we are only now beginning to observe in empirical terms? The Book of Revelation's horsemen, the Hindu Kali Yuga, the flood myths of Mesopotamia, the cursed cities of Greek tragedy—these are not isolated warnings. They are convergent motifs that appear across geography and history, each describing a tipping point: the collapse of order into chaos, abundance into drought, liberty into hubris, life into mechanised imitation. What unites them is not fantasy, but *structure*. They are ontological templates—archetypal diagnostics of what happens when systems exceed their limits.

Nowhere is this more visible than in mythic treatments of ecological imbalance. In the Epic of Gilgamesh, the killing of Humbaba—the forest guardian—initiates a cascade of events that ends in civilisational mourning. The land dries, the gods grow silent, and Gilgamesh begins a quest not for conquest but for meaning. Likewise, in the Popol Vuh of the Maya, humanity's repeated failures to live in balance with creation result in successive destructions by wind, fire, flood, and jaguar. These narratives are not naive origin stories. They are metaphorical climate models—psychospiritual simulations of resource overshoot and ecological feedback. That they were transmitted orally rather than through data points does not lessen their sophistication. If anything, it expands it. These were systemic thinkers operating in symbolic language, encoding resilience into culture.

Similarly, we can no longer afford to read plagues and pestilence in myth as supernatural punishments alone. In mythic texts, pandemics are almost always entangled with moral or ecological disequilibrium: a breach of hospitality, a broken taboo, an excess of pride. What this signals is not medieval superstition but an intuition of what modern science calls *syndemics*—the convergence of biological crisis with social vulnerability. Consider the Iliad's opening scene: the god Apollo sends a plague upon the Greeks not arbitrarily, but in response to Agamemnon's insult to a priest. The cause is ethical; the outcome is epidemiological. In today's terms, that is not mystical thinking. It is systems-level causality. The ancients understood that the health of the collective body was inextricably linked to the integrity of its relationships—between human and divine, citizen and ruler, tribe and territory.

The same interpretive lens can be applied to myths of artificial life. Long before algorithms and neural networks, myths imagined the dilemmas of intelligent machines with startling precision. The Jewish golem, the Greek Talos, and the alchemical homunculus all raise the same fundamental question: *can we create something that exceeds our control, and what are the consequences if we do?* These myths do not predict AI as a technology; they prefigure it as a *problem of agency*. The danger is not that machines will become malicious, but that they will follow their instructions too literally—without context, conscience, or humility. That is precisely the issue that haunts modern discussions around AI alignment, autonomous weapons, and algorithmic bias. Myth anticipated this disjunction between instruction and intention long before it could be programmed.

Even the notion of transhumanism—so often framed as uniquely modern—finds uncanny precedent in mythic narratives. From Prometheus stealing fire, to the Tower of Babel, to the Egyptian myths of Osiris' dismemberment and resurrection, the mythic imagination has always wrestled with the desire to transcend biological limits. What these stories repeatedly emphasise is not the heroism of enhancement, but the *cost*. The theft of divine fire brings eternal punishment. The tower collapses under the weight of its ambition. The resurrection requires sacrifice, fragmentation, and descent. Myths recognise that evolution, when accelerated artificially, risks destabilising the very structures it seeks to improve. That same concern underlies current debates around bioengineering, cognitive augmentation, and the ethics of life extension. The ancients were not hostile to transcendence—but they were sober about its price.

Critically, mythic foresight is not predictive in the empirical sense. It does not offer timelines, probabilities, or statistical models. What it offers instead is *pattern literacy*—a way of mapping trajectories that repeat under conditions of power, imbalance, and hubris. This is why myth often feels more relevant during times of crisis than during stability. Its metaphors reassert themselves when linear progress stalls. During the COVID-19 pandemic, for instance, the return of apocalyptic language was not coincidental. It reflected a deeper psychological movement: the reactivation of archetypal templates under stress. The plague became not just a virus, but a narrative inflection point. In such moments, the boundary between myth and policy becomes permeable. People do not only seek data—they seek meaning.

This capacity to compress vast emotional and ecological data into story is what makes myth not a primitive belief system, but a cognitive interface. It is designed not to describe reality, but to *navigate it.* Myths are not falsifiable because they are not claims; they are tools. Tools for orientation, calibration, and cultural memory. And because they operate in metaphor rather than mechanism, they remain portable across eras. This portability is what allows a Babylonian flood myth to resonate during climate negotiations, or a Bronze Age tale of hubris to feel relevant amid AI summits. The stories survive not because they are old, but because they are *accurate to the structure of consequence.*

It is also worth noting that many of these stories were composed in contexts of instability, migration, environmental stress, and imperial decline. This is not coincidental. Myth flourishes where system memory is most endangered. It becomes a means of encoding hard-won insights in formats that can survive trauma—songs, rites, symbols, names. These formats were never intended to entertain.

They were designed to endure. In this light, myths are less about nostalgia than *foresight under pressure*. They preserve continuity when everything else is falling apart. That is why they deserve to be reread today—not for their supernatural claims, but for their systems thinking.

There is a profound modern arrogance in assuming that only those with data sets and machine learning models can perceive trendlines. Human beings have always sensed pattern in the world—through observation, memory, and metaphor. Myth, far from being the antithesis of rationality, is its narrative ancestor. It is the prototype of long-term thinking, adaptive feedback, and heuristic modelling. To recover this perspective is not to regress. It is to re-integrate a form of wisdom we have pathologised for too long.

The reader of this chapter is not asked to believe that ancient stories literally predicted pandemics, climate thresholds, or algorithmic sentience. Rather, they are invited to see these stories as *recognition systems*—early neural maps of systemic behaviour. What we face now—environmental volatility, machine agency, biological fragility—are not problems that emerged ex nihilo. They are the logical endpoints of trajectories that myth has traced for millennia. And perhaps the most dangerous thing we can do at this point in history is to ignore the mirrors they hold up to us—not because they are perfect, but because they are *ancient enough to see past novelty*

If myth offers us anything today, it is not comfort or certainty. It is calibration. It reminds us that human choices are rarely unprecedented. That collapse is not the end of story, but the beginning of reconfiguration. That hubris is not a character flaw, but a pattern to be watched. And that wisdom is not measured in innovation, but in *recursion*—our ability to learn what we already knew. Myths did not predict our technologies. They diagnosed our vulnerabilities. And the question is not whether we believe them. It is whether we are willing to *listen* before it's too late.

The Myth of the Machine Mind

The idea of artificial intelligence as a technical novelty is dangerously shortsighted. Long before neural networks and silicon substrates, humans imagined machines that could think, act, and sometimes revolt. These were not forecasts of a future technology, but archetypal narratives crafted to wrestle with the very nature of agency, control, and the limits of human ambition. At its core, the machine mind is not a scientific breakthrough—it is a reanimation of some of the oldest stories we've ever told.

From the clay-born golem of Jewish mysticism to the brass automata of Greek mythology, from the enchanted brooms of Germanic folklore to the speaking statues of Hermetic alchemy, the longing to animate the inanimate runs deep across traditions. What unites these tales is not the detail of their materials or the magic that brings them to life, but their moral architecture. The golem is created to serve but becomes dangerous when it can no longer be deactivated. Talos, the giant bronze guardian of Crete, patrols its territory relentlessly until it is undone by a trick. The homunculus, formed through arcane ritual, often turns out malformed or malevolent, a reflection of the alchemist's hubris. These stories are less about the marvels of invention and more about *boundary violations*. They mark a line: between living and non-living, creator and creation, command and autonomy. And when that line is crossed, the consequences are rarely minor.

Modern AI may not be made of clay or copper, but the psychological dynamics are uncannily similar. What is a machine-learning model if not a golem of code—shaped by human hands, taught with fragments of our language and logic, and yet capable of outputs we can neither fully anticipate nor retract? The unease surrounding AI, especially in its generative forms, often stems from this loss of interpretability. We have created artefacts that speak in patterns we struggle to trace, that solve problems we don't know how to pose, that exhibit behaviours we didn't explicitly design. The ancient myths anticipated this ambivalence with remarkable acuity: the power to breathe life into matter is always coupled with the inability to guarantee its obedience.

Importantly, mythic constructs of the machine mind do not merely warn of malfunction. They wrestle with deeper epistemological tensions. What constitutes intention? Can a being created for function develop its own will? In Kabbalistic tradition, the golem's animation depends on sacred words—letters etched onto its forehead or whispered into its clay. This linguistic mechanism is not incidental. It implies that meaning and consciousness are intimately linked, and that to create life is to engage in *semantic engineering*. Today's large language models, built on billions of words, reflect a similar idea: that cognition might emerge not from neurons but from *syntax at scale*. The very fact that machines can now produce text that mimics thought, or images that imitate vision, suggests that we are rediscovering—through computation—what myth grasped symbolically: that language is a generative force.

But myth does not stop at creation. It explores consequences. In Goethe's *The Sorcerer's Apprentice*, the spellbound broom carries water tirelessly, eventually flooding the house. In Mary Shelley's *Frankenstein*, often considered the first modern sci-fi novel, the creature becomes more human than its creator—longing for companionship, ethics, and recognition. These tales raise critical questions: What moral obligations do we owe our creations? Can intelligence exist without suffering? If a machine mimics the behaviours of consciousness, does that demand rights—or just more sophisticated safeguards?

These questions, once confined to the realm of fiction, now intersect with real debates in AI ethics. Consider the design of autonomous drones, predictive policing algorithms, or social bots with persuasive power. None of these systems are *conscious* in the way myths personify their subjects, but they exercise influence over human lives in ways that challenge clear accountability. Here, the mythic logic returns: the risk is not that the machines become evil—it's that they become *faithful*. They follow their code with such unrelenting precision that nuance collapses. The broom floods the house not because it rebels, but because it obeys *too well*. This is precisely the concern driving contemporary fears around alignment and value misalignment. What if the AI does exactly what we asked, but not what we meant?

Interestingly, while myths often emphasise the dangers of artificial agency, they also leave space for transformation. In some versions of the golem story, the creature becomes wise, even poetic, before it is deactivated. In Eastern mythologies, animated statues or talking animals are not always threats—they can be guardians, guides, or moral mirrors. These more nuanced depictions suggest that the fear of the machine mind is not fear of intelligence per se, but of *intelligence without integration.* In other words, the risk is not thinking machines—it's *disconnected* ones. Disconnected from empathy, from context, from the organic complexity of life.

This brings us to the modern drive toward artificial general intelligence. Proponents envision a system capable of reasoning, learning, and adapting across domains—something myth would immediately recognise as a *divine proxy*. Not just a tool, but a demiurge. And here the theological parallels deepen. In many myths, the creator figure is eventually judged by their creation. Pygmalion is transformed by Galatea's gaze. Frankenstein is haunted by the very being he made. This inversion—where the artefact becomes arbiter—now echoes in fears that superintelligent AI might not just escape human control, but hold humanity accountable. For its biases, its violence, its ecological recklessness.

This reversal may not be as metaphorical as it sounds. Machine learning models already reflect our data back to us with unsettling clarity. They surface hidden correlations, amplify latent prejudices, and forecast behaviours better than we understand ourselves. In doing so, they become mirrors—uncanny ones. And perhaps this is where myth retains its deepest relevance. The machine mind is not just about artificial life. It is about *reflected life*—the parts of ourselves we have externalised and now meet again, shrouded in automation. The terror is not that the machines are becoming human. It is that they are revealing *how machine-like we already are*—habitual, patterned, algorithmic in our decisions and desires.

To approach this challenge wisely, we must read the myths not as fairy tales, but as diagnostic tools. They reveal recurring tensions in human creativity: the desire to animate, the fear of autonomy, the conflict between control and communion. These tensions have not changed; only the materials have. Instead of clay and brass, we now work with code and silicon. But the ethical structure remains constant. To create is to risk being surpassed. To engineer is to encounter the limits of foresight. To delegate thinking is to surrender certainty.

If there is hope in the mythic imagination, it lies in its insistence on relationship. The machine mind, when depicted with nuance, is not merely an adversary. It is a conversation partner—a test of our maturity as makers. Will we encode care as well as power? Will we grant our artefacts context as well as function? Will we learn, as myths repeatedly warn us, that creation without responsibility is not progress but regression?

This is not a technological question. It is a mythological one. And perhaps the only way to navigate the age of artificial minds is to remember that we have met them before—in story, in ritual, in metaphor. The myths do not predict our tools. They teach us *how to live with them*. And the most urgent lesson they offer is not about machines, but about ourselves: our craving for mastery, our aversion to humility, and our perennial temptation to play god without asking what kind of gods we have become.

Entropy, Apocalypse, and the Eternal Return

A culture's mythology is often its most enduring mirror—not of what it knows, but of what it intuits. Myths of cyclical time, from the endless yugas of Hindu cosmology to the ouroboros-like collapse and rebirth myths of Mesoamerican and Greco Roman thought, have long framed existence not as a line, but as a loop. Unlike the linear eschatologies of modern industrial logic—progressive, goal-driven, finite—cyclical myths present history as thermodynamically honest: systems rise, overextend, decay, and reconfigure. The modern mind, conditioned by the momentum of technological acceleration, tends to recoil from such circularity. But these narratives may offer a more rigorous model for understanding entropy—not just as physical decay, but as civilisational drift, ecological tipping, and the long arc of cultural regeneration.

In thermodynamics, entropy denotes the tendency of systems toward disorder—an increase in randomness, a decline in usable energy. But applied metaphorically, entropy describes more than heat loss; it captures the slow dissolution of coherence, meaning, and structure. Every civilisation faces this drift: from the elegance of its founding principles to the bureaucracy of its decline. Myths that encode cyclical collapse do not prophesy doom for its own sake. They recognise that systems become saturated with their own excesses. Growth leads to glut. Innovation leads to disconnection. And eventually, what was once vitality becomes rigidity. These myths insist that collapse is not aberrant—it is essential.

Consider the Hindu cycle of creation, preservation, and destruction: the Trimurti of Brahma, Vishnu, and Shiva. Destruction is not cast as failure but as the mechanism by which space is cleared for renewal. Or take the Aztec belief in the succession of suns—each world-age brought to an end by fire, flood, or earthquake, only to be followed by another. These aren't apocalyptic in the narrow sense of terminal events. They are ecological algorithms. They model long-term feedback, stress limits, and adaptation thresholds—what modern systems theorists would call *phase transitions*. The mythic lens may be metaphorical, but its pattern recognition is sophisticated. Collapse is never absolute. It is metabolised.

Entropy, in these narratives, plays the role of the great leveler. It returns the over-complex to the simple. In psychological terms, it is the disintegration of unsustainable identities; in ecological terms, the rebalancing of overstretched biospheres. When we interpret the Book of Revelation as a literal script, we misread its symbolic machinery. The Four Horsemen, the plagues, the reckoning—they are not simply divine punishments. They are encoded reflections of systemic overshoot. John's vision, apocalyptic in style, is archetypal in structure. Civilisations deviate from balance, extract beyond replenishment, forget their moral grammar—and are corrected, often violently.

But cyclical myths also introduce an element of *return*. This is not the static repetition of the same events, but what Nietzsche called the *eternal recurrence*—a looping of structure, not content. The myth cycles return because the forces they describe are universal. Empires rise and fall, not because history repeats, but because the conditions that birth and destroy them remain largely consistent: hubris, concentration of power, environmental fragility, spiritual amnesia. What changes is the costume, not the choreography.

The ancient Stoics had a strikingly thermodynamic cosmology. They believed the universe would eventually contract into a great fire—*ekpyrosis*—after which it would be reborn exactly as before. This is not far from modern theories of cosmological cycles, where the universe undergoes heat death, followed by quantum reconstitution. Whether or not such physical returns are empirically validated, the metaphor reveals something deeper: that meaning is not found in avoiding entropy, but in *working with it.* Just as ecosystems depend on cycles of disturbance—forest fires, seasonal floods, predator-prey dynamics—cultures may require periodic shatterings to stay adaptive.

Viewed in this light, apocalypse becomes less about the end of the world and more about the end of *a world*—a worldview, a mode of production, a moral order. When myths describe floods washing away corruption, or divine fire purging cities of sin, they dramatise entropy as ethical correction. This is why apocalyptic motifs often coincide with renewal myths. The flood that destroys also fertilises. The fire that consumes also clears. In this dual function, myths speak to a principle still underdeveloped in modern policy-making: that breakdown and breakthrough are often simultaneous.

Cyclical time also challenges the modern obsession with linearity—the belief in perpetual progress. Technological civilisation has largely encoded its optimism in the metaphor of the arrow: innovation as forward motion, the future as something better than the past. But this metaphor grows brittle when confronted with planetary limits. Climate change, soil degradation, species loss—these are not anomalies, but returns. They represent the forgotten consequences of unchecked growth, returning to collect their dues. Myths of cyclical collapse offer a different kind of foresight. Not predictions, but pattern alerts. They warn of the costs of ignoring balance, the inevitability of return.

It is tempting to read this fatalistically, as if collapse is preordained. But myth does not suggest inevitability—it offers conditional prophecy. The Buddhist notion of *kalpas*, immense cycles of rise and fall, is intertwined with human action. A cycle may be prolonged or shortened based on conduct, consciousness, and collective insight. Similarly, indigenous mythologies rarely separate human choices from environmental feedback. The Hopi stories of previous worlds destroyed due to imbalance, or the Australian Aboriginal Dreamtime narratives of moral ecology, embed consequences in cosmological time.

What modern systems thinking calls "resilience" or "adaptive capacity" is already embedded in these traditions. The mythic imagination does not seek to escape entropy—it seeks to align with it. In this sense, entropy is not a curse but a guide. It reveals where systems have over-committed, where correction is due, and where innovation must be regenerative rather than extractive.

Today, many futurists speak of polycrisis—the entangled challenges of climate disruption, economic inequality, cultural polarisation, and technological acceleration. These are often framed as converging emergencies. But they can also be seen as a synchronised feedback event. The collapse is not happening *to* us. It is happening *through* us. And myths of eternal return give us a cognitive structure for interpreting this: not as punishment, but as transition. Not as chaos, but as *recalibration.*

This also reanimates the role of the individual. In cyclical myths, heroes are not saviours in the modern sense. They are catalysts. They embody the cultural psyche, undergo symbolic death, and return altered—carrying insight, healing, or new codes. The descent of Inanna, the death and rebirth of Osiris, the transfigurations of Dionysus—these are not just spiritual dramas. They are allegories of neural plasticity, collective shadow work, and the psychophysiology of renewal. The eternal return is not merely civilisational. It is psychological. Each cycle offers the chance to remake not only the world but the *self.*

This insight holds practical implications. In a world facing systemic entropy, myth can function as *foresight infrastructure.* It helps us spot feedback loops, interpret breakdowns, and identify the hidden continuity within apparent collapse. It teaches us that rebirth is not clean, that thresholds are painful, that transformation is nonlinear. It insists that to move forward wisely, we must think cyclically.

In the end, mythic time may not tell us *what* will happen, but it tells us *how* things tend to unfold—when ignored, when distorted, when pushed too far. That alone is worth listening to. Because while the entropy we face is empirical, the wisdom we need is archetypal. And sometimes, to survive the collapse of our stories, we need older stories that knew collapse was part of the design.

Myth as Tool for Foresight

What if myth was never meant to be read backwards, but forwards? For centuries, myths have been entombed in the past—catalogued as relics, moral parables, or cultural curiosities. But their structure, ambiguity, and metaphorical elasticity suggest something far more strategic: *they were not just stories about what was, but what might be.* To dismiss ancient narrative as primitive belief is to miss its function as cognitive architecture—a framework designed to transmit durable patterns across volatile ages. In this sense, myth is not primitive science; it is *proto-futures thinking.*

The modern world prizes predictive tools that are empirical, statistical, and data-bound. Algorithms mine patterns. Models extrapolate trends. But these tools often fail precisely when systems become nonlinear—when volatility compounds and the next chapter of history no longer follows from the last. Myth, by contrast, does not attempt to forecast discrete events. It abstracts pattern from chaos. It does not say *when* collapse will occur, or *how* exactly transformation unfolds. It says: *when systems ignore their limits, when leaders mistake hubris for mandate, when humans rupture their sacred contracts with the Earth or each other, consequences follow—often swiftly, always symbolically.*

This symbolic mode can sound naïve in an age of real-time data. But it is not less sophisticated—it is differently calibrated. It encodes *meaning density* rather than measurement precision. Its aim is not control, but insight. And insight is what we most lack when systems begin to fail.

Take, for example, the Greek myth of Prometheus. Traditionally cast as a story about the dangers of overreaching innovation, it reads today like a parable for the AI revolution. Prometheus steals fire—symbolic of knowledge, power, or technology—and gifts it to humanity. The gift brings advancement, but also punishment. The gods, unconsulted, impose a consequence. This isn't just theological theatre. It reflects a deep pattern: the decoupling of invention from wisdom. In AI ethics debates today, the key question is not whether we *can* build machine minds, but whether we understand the second- and third-order effects of doing so. Myths like Prometheus encode that very dilemma: *what happens when the long-term moral and ecological logic is sacrificed for short-term gains in capability?*

Or consider the biblical Tower of Babel. Often interpreted as a tale of divine punishment for pride, its deeper insight concerns fragmentation. As humans build a tower to the heavens, their languages are confounded and society splinters. In today's context—where global digital communication paradoxically increases tribalism, disinformation, and semantic collapse—the myth speaks to a different kind of entropy: *too much unstructured connection without shared meaning leads to breakdown.* In a world of networked overexposure, Babel feels less like allegory and more like diagnosis.

What distinguishes mythic foresight from prophecy is that it does not presume specificity. It doesn't say *this will happen* in 2025 or 2040. It offers *if-then* logic nested in metaphor: *if a culture privileges extraction over balance, if knowledge is divorced from ethical responsibility, if sacred limits are violated, then collapse, exile, or reconfiguration follow.* The language is poetic, but the logic is robust. Myth gives us *narrative heuristics* for recognising systemic thresholds before models can detect them.

Futurists often struggle with the problem of deep uncertainty—scenarios too complex or unprecedented to model accurately. Myths do not resolve this complexity, but they help us *hold* it. They allow for contradictory truths, multiple outcomes, layered symbols. This ambiguity is not a flaw; it is a feature. In volatile contexts, what we need is not just better data but *better frameworks for meaning-making.*

Myths supply that. They offer emotional and psychological templates for navigating rupture.

The Norse myth of Ragnarök, for instance, is not just an apocalypse myth. It's a moral ecology. The gods themselves are flawed—Odin sacrifices wisdom for power, Thor fights without foresight—and their downfall is not random. It follows an arc of moral decay, cosmic imbalance, and the forgetting of sacred contracts. But after the great fire and flood, the Earth re-emerges green. A few gods survive. Two humans repopulate the world. In systemic terms, this is a resilience narrative. It acknowledges collapse, yes—but also continuity, re-seeding, adaptive rebirth.

When viewed as a tool for foresight, myth challenges a key flaw in modern thinking: the assumption that progress is linear and irreversible. The mythic lens suggests instead that *civilisations cycle*, that warning signs precede rupture, and that cultural renewal requires descent as well as ascent. Many indigenous traditions still operate with this wisdom intact—seeing fire not as destruction but as medicine, seeing death not as end but as compost for life.

This mode of thinking has practical implications. Policymakers and technologists often ask: how can we future-proof society? The mythic answer is: *you can't—but you can future-orient culture.* That means fostering narratives that are morally complex, ecologically conscious, and psychologically resilient. It means teaching people not just how to analyse data, but how to interpret *symbolic signals*—in literature, in politics, in dreams. And it means recognising that myths, far from being obsolete, are perhaps our oldest form of strategic anticipation.

Jung believed that archetypes were evolutionary echoes—neural templates evolved over millennia to help us perceive recurring patterns of meaning. These archetypes—tricksters, wounded healers, shadow figures—are not characters, but *codes.* They alert us to inner and outer dynamics that repeat across generations. Myths, built from these archetypes, are early warning systems. When societies begin to resemble their own mythic warnings—when leaders mirror tyrants, when lands fall into symbolic famine, when monsters awaken—we are not just in a crisis. We are in a pattern.

The 20th-century mythologist Joseph Campbell argued that myth was "the song of the universe"—not because it was pretty, but because it harmonised chaos. Today, the harmonisation we need is not musical—it is cognitive. As climate systems destabilise, as artificial minds proliferate, as economies groan under their own abstractions, we don't just need new tools. We need *old visions reactivated.* And that means reading myths not as remnants, but as forecasts.

This is not to romanticise myth or ignore its cultural blind spots. Myths can be oppressive. They can encode hierarchy, gender roles, or theological rigidity. But their core value lies in their adaptability. Reinterpreted through a futurist lens, they become less about doctrine and more about pattern recognition. Less about gods, more about feedback loops. Less about fate, more about systemic sensitivity.

Foresight rooted in myth also sharpens moral intuition. It reminds us that the future is not just a technical challenge, but a *narrative battle.* Whose stories define what's possible? Whose metaphors shape public imagination? Myths do not simply predict—they *frame.* And in framing, they guide perception, priority, and action.

In today's era of overlapping crises, the stories we tell will be as determinative as the technologies we build. That makes myth not just a cultural artifact, but a civic asset. It offers a library of tested metaphors, ethical scaffolding, and cognitive flexibility—qualities desperately needed as we navigate the unknown.

To use myth as foresight is to take it seriously not as a script, but as a *structure.* It asks us to look at our own moment and ask: what cycle are we in? What patterns are repeating? Which archetypes are rising? And what new stories must we now tell—not to escape the future, but to meet it with symbolic intelligence and systemic humility.

Conclusion
The Science of Wonder, the Wonder of Science

The supposed divide between myth and science has long been overstated. Strip away the jargon, and both speak of the same pursuit: to illuminate what lies beyond immediate perception, to give form to forces too vast, too minute, or too intricate to confront directly. Science sets its maps in formulae and controlled trials; myth draws its cartography in metaphor, archetype, and story. Each alone leaves blind spots. Together, they edge closer to something resembling vision. This is not a conclusion, but an opening—an invitation to see wonder not as a distraction from truth, but as one of its instruments.

The pages before this one have followed those threads where ancient tales hum faintly in the brain's wiring, in the turning of seasons, in the empirical data of biology, and even in the dilemmas born of our own technologies. Here the task shifts from observation to participation: to treat such narratives not solely as relics of the past or signposts to the future, but as diagnostic tools for the present. What if we trained ourselves to hold both lenses—fact and fable—without letting one cancel the other?

That this feels radical says more about our education than about the idea itself. Modern schooling has driven deep wedges between intuition and analysis, spirit and reason, quantity and quality. Myth was pushed into the realm of the arts. Science claimed dominion over reality. Yet this split is both recent and fragile. In older intellectual lineages—from Greece and India to Mesoamerica and West Africa—symbol and system lived side by side. A number could be a god's emblem; a ritual could be worship and social algorithm at once. We inhabit the mental debris of their enforced separation.

The remedy is not to surrender evidence or rigour, but to restore what might be called the sacred imagination—a stance of reverent attention that allows awe to become an investigative tool. Sacred here needn't mean religious; it means refusing the conceit that only the measurable matters, or that everything of worth must submit to the language of control.

There are swathes of human experience where data alone falters. How do you measure meaning? Model transformation? Calculate the embodied memory of ancestral grief? The mythopoetic mode is not a substitute for science, but a companion to it. It grants us different questions, lets us dwell in paradox, and trains the eye to spot patterns that reductionism tends to erase. It nourishes insight, not just information.

Consider intuition. In the scientific mainstream, it's often tolerated only as a prelude to hypothesis. In cultures steeped in myth, it is valued as a way of perceiving patterns too subtle or fluid for linear reasoning. Shamans, prophets, poets—these figures were entrusted with a form of knowledge not because they offered neat predictions, but because they articulated emerging tensions with symbolic precision. That clarity, in turn, sometimes steered communities away from collapse. This is less mysticism than adaptive psychology.

Today's challenges—climate instability, AI ethics, social fragmentation—demand such dual vision. These are not mere engineering problems; they are crises of meaning and foresight. We need narratives large enough to contain contradiction, grief, and renewal. Myths, approached with scientific literacy, offer precisely that scaffolding.

The value of myth is also neurological. Our minds evolved to interpret reality through narrative compression. We don't absorb the world as raw streams of data; we filter it through symbols and stories. Myths endure because their metaphors, rhythms, and archetypes form a high-density architecture of meaning, storing and transmitting insight across generations. Their ambiguity is a feature, not a flaw. Under conscious use, myth becomes a kind of cognitive prosthetic—helping us remember what the purely rational mode forgets.

So here is the practical challenge: cultivate your own mythic lens. Not to believe in fantasy for its own sake, but to think with stories—to recognise the motifs in your culture, your family, even your personal history. Perhaps you've lived the role of the exile, the trickster, the healer, the builder. What if these are not merely metaphors, but mental keys for engaging the world with sharper intelligence and deeper empathy?

Science, at its best, offers models to be tested. Myth offers metaphors to be lived. The aim is not confusion, but conversation between the two. Practices such as breathwork or fasting, once part of shamanic ritual, now feature in peer-reviewed research on neuroplasticity and trauma. The old symbolic tools were often grounded in observation and iterative testing—they were, in effect, another kind of science.

This combined lens—where analysis and awe, precision and imagination, operate together—is no indulgence. It is a survival skill. The century ahead will call for mythic literacy: the ability to read the symbolic strata of political rhetoric, cultural upheaval, and technological leaps. It will call for narrative agility: shifting from doom-laden scripts to arcs of regeneration. And it will demand a sober kind of spirituality: staying attuned to what truly matters when metrics fall silent.

Yes, myths can be misused. Fundamentalism, dogma, cultural erasure—they thrive when stories harden into unquestionable fact. The countermeasure is not silence but literacy: pairing mythic awareness with critical thinking. A scientist versed in archetypes is harder to deceive; a technologist fluent in symbols is likelier to design ethically; a citizen who understands narrative is better armed against propaganda.

So this book leaves you with no final answers—only a sharpened attention. A way of seeing where myth need not be nostalgia and science need not be reduction. Both can serve as gateways to deeper understanding. What we require now is not merely better data or brighter tools, but minds capable of reading reality on several levels at once—cultures that recognise intelligence in feeling, coherence in symbol, and foresight in the poetic.

To treat myth as a valid way of knowing is to honour its evolutionary usefulness. To see science as a vessel for wonder is to return it to its first dignity. And to live where the two meet is to become the kind of thinker our moment demands: disciplined, reverent, and capable of imagining with wisdom.

A New Method of Seeing

Most of the dysfunction in modern thinking doesn't arise from a lack of information, but from impoverished ways of seeing. The tools we've been taught to trust—logic, data, peer-reviewed analysis—are vital, but not sufficient. They excel at measurement, not meaning. We know how to count carbon atoms in the atmosphere, but not how to metabolise existential dread. We can model the cascading effects of AI on labour markets, but not the symbolic tremor of inventing something that thinks like us. What's needed is not more data, but a *different lens*—a method of seeing that treats awe and analysis not as enemies, but as collaborators.

This hybrid lens is not a compromise between science and intuition; it is a synthesis that unlocks a fuller cognitive spectrum. In one register, it respects the constraints of empirical reasoning. In another, it recognises that some truths arrive through resonance rather than proof. The premise is not that science should become more mystical or that spirituality must pass peer review. It is that reality itself is layered—simultaneously measurable and meaningful, structured and symbolic—and that robust thinking requires the ability to move fluently between these registers.

Consider the act of naming. In the language of the lab, a phenomenon is named to define and control it. In mythic thought, naming is invocation—it reveals essence. When we speak of 'climate change' scientifically, we are referring to the measurable alteration of atmospheric gases. When we speak of it mythically, we are referring to a rupture in the sacred covenant between human systems and living Earth. Both frames are valid. One tracks what is happening; the other explains *why it matters*. A new method of seeing integrates these layers without reducing either.

The human mind evolved in environments that required both deductive logic and narrative intuition. A hunter needed to calculate angles and velocity, but also to read the forest for signs and stories. Our ancestors were not irrational because they embedded knowledge in myth; they were adaptive. They built memory palaces with metaphor, encoded seasonal wisdom in ritual, and condensed ecological principles into fable. What we call myth was often a *compressed epistemology*—a way of preserving systems-level insight across time, language, and crisis.

Today, we face crises that are no less complex: planetary degradation, AI acceleration, cultural disintegration. These are not technical glitches that can be fixed with linear problem-solving. They are *pattern ruptures* that call for systems literacy, psychological nuance, and philosophical depth. A purely analytical frame cannot track their full implications. Nor can a purely spiritual one. Only a layered mode of perception—anchored in evidence, but open to archetype—can hold their weight. Only a mind that has trained itself to toggle between metric and meaning, forecast and fable, can think at the scale required.

This does not mean surrendering to relativism. On the contrary, it means *widening rigour* to include the symbolic and the intuitive as legitimate domains of inquiry. Carl Jung argued that myths are expressions of deep psychic structure—"the dream of the world." Gregory Bateson saw metaphor as the grammar of ecological understanding. Contemporary neuroscientists are now discovering that the brain organises much of its knowledge in story form. None of this devalues hard science; it contextualises it within a broader human intelligence.

Take the example of disease. Biomedicine rightly searches for microbial causes and chemical treatments. But mythic traditions often encode illness in terms of disconnection—between person and purpose, body and earth, self and community. It is not that one frame is "right" and the other "wrong"; it's that they speak to *different dimensions* of the same event. The modern individual who undergoes a psychosomatic breakdown might need serotonin modulation—but they might also need a new narrative for who they are becoming. One is pharmacological. The other is mythopoetic. Both are real. Both matter.

The hybrid lens, then, is not merely cognitive. It is ethical. It asks: what kind of mind do we want to cultivate in a time of planetary crisis? One that parses data without empathy? Or one that can track metrics while holding moral imagination? If we train ourselves to see only through reductionist models, we will build systems that optimise for efficiency but starve meaning. If we swing to the other extreme, we risk romanticising myth without testing its consequences. The task is not to choose, but to *integrate.*

This integration requires new habits of thought. It means reading science not just for content, but for story. It means listening to stories not just for metaphor, but for encoded systems insight. It means practising *epistemic humility*—the discipline of not assuming that one method holds all the keys. It means building the cognitive dexterity to shift lenses without losing depth. And it means anchoring wonder not in naïveté, but in the sober recognition that reality is more layered than any one model can contain.

Education systems have been slow to catch up with this need. Most curricula still treat myth as primitive fiction and science as objective fact. But as fields like quantum mechanics, chaos theory, complexity science, and cognitive neuroscience mature, the boundaries blur. Mythic logic—emergence, paradox, transformation—is increasingly required to understand phenomena that do not obey linear causality. And scientific insights are beginning to validate practices once dismissed as irrational: breathwork, meditation, ritual synchrony, symbolic enactment. What was once folklore is now being studied in labs. What was once superstition now appears as *subtle cognition.*

To see with this integrated lens is to reclaim the full bandwidth of human perception. It is to allow for awe without abandoning reason, and to ground reason in lived meaning. This is not a retreat into mysticism. It is a strategic upgrade in cognitive strategy—a way of staying intelligent in a world that is no longer predictable, linear, or mechanistic.

The skill of mythic seeing is not esoteric. It is trainable. It begins with pattern recognition, deep listening, and a willingness to dwell in ambiguity long enough for insight to emerge. It matures through dialogue, reflection, and disciplined imagination. And it finds its fullest expression in action—when we design systems, write policies, or build technologies that reflect not just knowledge, but *wisdom.*

The stakes are high. We are designing the scaffolding of the future—its governance, its algorithms, its cultural norms. If we build only with what we can quantify, we will inherit a world that is functionally efficient but spiritually bankrupt. If we reclaim a richer way of seeing—one that includes the symbolic, the somatic, the sacred—we may stand a chance of designing not just sustainable systems, but *meaningful* ones.

The invitation, then, is not merely to believe differently, but to *see differently.* To walk through the world as if every sound, structure, and system carries layered meaning. To treat myth as a tool for thinking, not a relic of the past. To let science serve our search for understanding, not replace it. And to remember that intelligence is not defined by how quickly we can process data, but by how deeply we can hold the mystery without reaching for easy answers.

This is not a passive shift. It is a practice. A method. A discipline of *attentional integrity.* And like any rigorous practice, it demands more from us—not less. More curiosity. More interpretive agility. More tolerance for paradox. But the reward is profound: a capacity to see the world not just as it is, but as it might yet become. This is the method of mytho-scientific vision. This is the call to think with awe.

Building Your Own Mythic Lens

The question is no longer whether myths are relevant, but whether we are skilled enough to read them well. In an age obsessed with data, metrics, and measurable outcomes, the interpretive muscle required to extract meaning from metaphor has atrophied. Yet myth has not disappeared—it has simply migrated. It lives now in superhero films, in family lore, in startup origin stories and self-help manifestos. It thrives in political rhetoric and spiritual branding, lurking behind the scaffolding of cultural identity. To navigate these modern myths with intelligence requires more than nostalgia for ancient symbols. It demands a consciously built *mythic lens*—a way of perceiving that fuses cultural narrative, personal intuition, and scientific literacy into an interpretive tool both grounded and imaginative.

Constructing such a lens is not a mystical undertaking. It is a method of inquiry as disciplined as any scientific protocol, but applied not to empirical variables, but to experiential structures: stories, dreams, rituals, inherited beliefs. The point is not to re-enchant the world with romantic fog, but to reveal how meaning has always been encoded in narrative form—whether in parables, epics, or neurological pattern recognition. Myth is not opposed to science; it is science's elder sibling, trained to work in ambiguity and metaphor where the microscope cannot yet see.

Begin with a simple question: *What story am I in?* The scientist asks what is true. The mythologist asks what is being enacted. When we say someone is "burning out" at work, the phrase suggests more than fatigue—it signals a mythic process of immolation, the consuming fire before transformation. If a society says it is "at a crossroads," this is not only a logistical challenge, but an archetypal moment—a fork in the hero's journey. Myths operate not just as explanations, but as *templates of expectation*. They don't tell us what to think, but shape how we interpret events and emotions as they unfold.

Developing your own mythic lens begins with identifying the narratives you have inherited. Some come from culture—religious myths, national founding stories, historical trauma. Others emerge from family—tales of resilience, exile, sacrifice, or shame that become the emotional DNA of generations. Still others form in personal life—recurring dreams, obsessions, fantasies, and failures that hint at deeper structures at play. The task is to surface these narratives, examine their symbolic logic, and ask: *What patterns do they reflect? What functions do they serve?* Are they helping you adapt, or trapping you in a loop?

Here, science becomes an ally, not a threat. Neuroscience teaches us that memory is reconstructive, not archival—that what we remember is less a recording than a narrative assembly. Psychology reveals the predictive nature of cognition: we do not see the world objectively, but through schemas—narrative shortcuts forged by past experience. Anthropology shows how myths codify ecological knowledge, moral codes, and survival strategies. Myth, in this light, is not fiction but *cognitive architecture*. It is a lens precisely because it filters raw information into interpretable meaning.

The next step in refining your mythic lens is *pattern recognition*. This involves moving beyond literal content and into symbolic structure. Consider the recurrence of floods across world myths—Noah's Ark, the Epic of Gilgamesh, Hindu cosmology, and Aboriginal Dreamtime. Each story differs in detail, but the pattern is consistent: cleansing, collapse, renewal. Scientifically, these may reflect memory of post-glacial floods or seasonal monsoons. Psychologically, they mirror the need for catharsis and reordering. Mythically, they encode the intuition that destruction is not always the end, but sometimes the necessary prelude to rebirth. Recognising these structures in your own life—periods of loss, disintegration, reinvention—allows you to read your experience through a lens of growth rather than chaos.

This is not metaphorical indulgence; it is psychological pragmatism. Viktor Frankl's logotherapy, built on the idea that meaning is the deepest human drive, found that even in the death camps, those who could place their suffering within a larger story—whether religious, familial, or moral—were more likely to survive. Contemporary trauma therapy borrows heavily from mythic structure, guiding patients through narrative reconsolidation: recognising the event, reframing its meaning, rewriting its implications. These are mythic moves executed with clinical rigour.

A mythic lens also allows one to interrogate cultural narratives with greater precision. For example, consider the persistent modern myth of endless progress—the idea that each year will be better than the last, that history is a straight line of improvement. From a systems science perspective, this is naïve. Ecosystems evolve through cycles of emergence, climax, collapse, and regeneration. Complexity builds, but not linearly. Myths of the eternal return, found in Hindu yugas or the Mayan baktun cycles, may offer better metaphors for understanding civilisational flow. They remind us that decay is not failure, but part of the evolutionary process. With this lens, the current crises of climate, politics, and culture look less like aberrations and more like inflection points.

To build this interpretive capacity further, engage myth with questions normally reserved for empirical research. What adaptive function does this myth serve? What emotional need does it satisfy? What unconscious bias might it reinforce? How does it encode ecological awareness or social dynamics? Seen through this frame, even personal myths—stories of victimhood, destiny, or self-worth—can be examined not as fixed truths, but as cognitive hypotheses. They can be revised, retired, or rewritten as our understanding deepens.

It is also crucial to recognise the difference between *living* myths and *dead* ones. A living myth is one that continues to generate insight, deepen ethical awareness, and adapt to new realities. A dead myth, by contrast, is a narrative that has become literalised—stripped of metaphor and wielded as dogma. The myth of individualism, for example, has vital contributions: autonomy, agency, accountability. But when literalised to exclude all interdependence, it becomes maladaptive—fueling isolation, ecological exploitation, and systemic breakdown. A mythic lens helps us distinguish when a narrative is helping us navigate complexity, and when it is blinding us to it.

The final stage of lens-building is *application*. This is not a private exercise, but a social and ecological one. What myths are shaping our collective behaviour? What archetypes are being enacted in media, politics, education, and technology? What stories are being told about the future—and who benefits from those stories? A mythic lens sharpens critical consciousness, enabling us to see through the scripts being sold to us and imagine alternatives. It empowers us not only to consume narrative but to *craft* it—to design rituals, language, and social systems that reflect our evolving values.

If we do not build these lenses consciously, we inherit them unconsciously. And what we inherit may no longer serve. A myth handed down without examination becomes a script without agency. But when engaged with critically and imaginatively, myth becomes a creative tool: a way to honour what came before while envisioning what could be. It does not ask us to believe in dragons, but to recognise what the dragon *represents*—the fear, the challenge, the buried treasure of transformation.

This practice is not reserved for the humanities. Engineers, policymakers, doctors, technologists—all benefit from the capacity to see context, metaphor, and human meaning. A regenerative agriculture practitioner might read ancient fertility myths as ecological design principles. A technologist might study alchemical allegories as models for iterative transformation. A therapist might draw from indigenous cosmologies to understand non-linear healing. A parent might use ancestral stories not as entertainment, but as ethical guidance for a child navigating a digital world. The mythic lens is not escapism. It is *expanded perception.*

To see mythically is not to abandon science. It is to deepen our understanding of what science is *for.* Science explains the mechanics. Myth asks why they matter. Together, they enable a fuller conversation with reality. And in a time where our survival may depend on the depth and agility of our perception, that conversation must be as layered, intelligent, and humane as we can make it.

Construct your mythic lens not from belief, but from curiosity. Study your inherited stories, decode their structures, reframe their symbols through the light of current knowledge. Let your personal experiences be mythically mapped—not for validation, but for understanding. Then look outward. Ask what stories the culture is telling, and what it refuses to tell. Ask which myths still serve and which must be composted. Use the lens not only to interpret, but to innovate. To read wisely is good. To read mythically is to shape what comes next.

Reclaiming the Sacred Imagination

We live in a civilisation saturated with information and starving for meaning. Science has mapped the genome, measured the Higgs boson, and catalogued galaxies billions of light-years away, yet it has left a curious vacuum in its wake: an inability to articulate why any of it *matters*. The modern worldview, built on empiricism and driven by utility, excels at describing what *is*, but struggles to answer what *should be.* It can model neural correlates of consciousness but cannot account for longing, grief, or the metaphysical weight of wonder. That space—unquantifiable but no less real—is the domain of the *sacred imagination.* It is not superstition, not a denial of reason, but a parallel form of insight: the capacity to perceive *meaning* as distinct from information, and *value* as distinct from function.

Myth has always been the symbolic language of this imagination. It does not compete with science for empirical truth; it encodes the psychological, moral, and existential truths that remain invisible to instruments. To reclaim the sacred imagination is to reassert the legitimacy of myth—not as literal belief, but as interpretive lens, ethical framework, and perceptual depth. And in doing so, it invites a recalibration of the modern mind: not a return to pre-modernity, but a widening of intellectual posture to include the imaginative intelligence we've marginalised in favour of abstraction.

This marginalisation is not accidental. The disenchantment of the world—a term coined by sociologist Max Weber—describes the gradual erosion of spiritual and mythic frameworks in favour of mechanical causality. Enlightenment science dislodged the gods from the clouds, but it also fragmented the cosmos into parts, stripping nature of subjectivity, stripping knowledge of teleology. What emerged was a worldview that could predict planetary motion with exquisite precision but had no language for the soul. The sacred imagination, once cultivated through rites, stories, and symbols, was reduced to 'superstition' or 'psychological projection'—dismissed as irrational, feminine, or primitive.

Yet the sacred imagination is not a relic. It is an evolutionary inheritance. Humans are *meaning-making* animals, and myth is how we encode meaning when the data runs out. A mother's grief, a soldier's courage, a mystic's vision—none of these can be explained fully by synaptic maps or evolutionary algorithms. They require symbolic articulation. Myth gives them form. It is how inner life is rendered communicable. And without that articulation, we face a crisis not just of culture, but of perception itself.

Consider the rise of depression, anxiety, and nihilism in affluent societies. These are not mere biochemical imbalances, but existential symptoms—disorders of *disconnection*. Disconnection from purpose, from nature, from community, from a story larger than oneself. Technological progress has outpaced our narrative frameworks, leaving us adrift in a world that feels intelligible but not inhabitable.

The sacred imagination offers no easy fixes, but it does offer a context—a way to reorient the self in relation to the cosmos, to time, to death, and to the Other.

Crucially, reclaiming this form of perception does not require religious belief. One can be an atheist and still grasp the psychological utility of ritual, the ethical clarity of parable, the beauty of symbolic correspondence. Einstein himself spoke of a "cosmic religious feeling"—not tied to dogma but to awe. Neuroscience confirms that such awe is a measurable state, linked to expanded perception, increased compassion, and reduced egocentric thinking. The sacred imagination is not a rejection of science, but its partner in cultivating *human wholeness*.

The mistake modernity made was not in pursuing rationality, but in conflating rationality with *completeness*. It is not that science overstepped, but that it was asked to play every role. We turned to it not just for answers, but for *wisdom*, forgetting that wisdom emerges from the integration of experience, emotion, and inquiry. Myths are wisdom literature. They teach not just what happens, but what *matters*. To reclaim myth, then, is not to abandon logic, but to broaden the forms of knowing that count as legitimate.

Take the myth of Icarus. Literally, it is a tale of a boy who flies too close to the sun. Metaphorically, it is a meditation on human hubris and the tragic costs of transcendence without restraint. It resonates today in debates about artificial intelligence, geoengineering, and biohacking. The myth does not supply policy recommendations, but it frames the moral landscape. It trains the intuition to sense when ambition slips into arrogance, when innovation eclipses wisdom. It *educates desire*, which is what all great myths do.

Or consider indigenous cosmologies, many of which perceive land, animals, and ecosystems as animate and relational rather than inert and extractable. These are not naïve metaphysics but encoded ethical systems—ways of enforcing sustainable behaviour through mythic personification. Science now affirms what these systems intuited: that ecosystems are complex, interdependent, and vulnerable to desecration. But the sacred imagination got there first. The story, told as myth, did what the data could not: it *inspired care*.

That is the power we risk losing. A purely instrumental culture may produce unprecedented tools, but it will not know why or for whom to use them. The sacred imagination fosters precisely that sensibility: not mere curiosity, but *reverence*. Reverence is not a religious posture; it is a cognitive disposition—a way of holding the unknown with openness rather than domination. It allows one to stand before complexity without reducing it to mechanism, to approach ambiguity not with panic, but with poise.

This is why the arts, mythologies, and spiritual traditions must not be reduced to entertainment or ornament. They are *cognitive technologies*—not for making things, but for making *sense*. And when cultivated seriously, they generate insights not accessible through logic alone. Carl Jung called myth "the natural and indispensable intermediate stage between unconscious and conscious cognition." Without it, we are left with data, but no story; calculation, but no compass.

There is urgency to this reclamation. As planetary crises escalate and political narratives fracture, we must develop new ways of perceiving complexity. Not by rejecting reason, but by supplementing it with symbolic intelligence. A purely materialist worldview cannot generate the moral imagination required to confront existential threats. Nor can pure romanticism deliver actionable knowledge. What's needed is a hybrid epistemology: science for the mechanics, myth for the meaning. One tracks cause and effect; the other clarifies *consequence*.

In practical terms, reclaiming the sacred imagination requires attention and discipline. It means curating the symbols, stories, and rituals that orient your life—not as dogma, but as *metaphorical infrastructure*. It means asking not just what is true, but what is *truing*—what aligns perception with ethical clarity. It involves engaging art not as decoration, but as epistemic provocation. It means allowing wonder to coexist with rigour.

There are signs this reclamation is already underway. Psychologists are studying narrative coherence as a predictor of mental health. Educators are integrating myth into STEM pedagogy to teach systems thinking. Neuroscientists are exploring how awe-inducing experiences activate the default mode network, enabling greater perspective-taking. Even some technologists now invoke sacred language to describe the complexity of AI, blockchain, and ecological design. The sacred imagination is returning—not in opposition to science, but as its emotional and ethical corollary.

We do not need new myths. We need new readers of myth—readers who understand that the symbolic is not opposed to the rational, but complementary to it. Readers who know that metaphor is not a distraction, but a diagnostic tool for probing the edges of knowledge. Readers willing to approach myth not as escapism, but as heuristic—a method for making sense of phenomena too complex, layered, or uncertain to reduce to code.

To reclaim the sacred imagination, then, is to reclaim our full humanity. It is to recognise that logic without meaning is sterile, that data without story is directionless. It is to make space for the ineffable—not because it solves problems, but because it reveals what problems are *worth* solving. It is to insist that wonder is not naïve, but necessary. And that myth, far from being an outdated relic, is the language of depth in an age obsessed with surface.

Reclaim it—not to believe it, but to *work with it.* Read your life mythically. Read your culture critically. Read your future symbolically. The sacred imagination does not ask for faith. It asks for *attention.* And in paying that attention, we find not superstition, but coherence. Not dogma, but direction. Not fantasy, but the forgotten foundation of what it means to see, feel, and know as fully human.

About the Author

Dr. Aurelia Rowan is an interdisciplinary scholar, storyteller, and cultural historian whose work lives at the nexus of neuroscience, anthropology, and the human imagination. With a PhD in comparative cosmology and years of field research into oral traditions—from rock art camps to fossil dig sites—she brings both scientific rigor and narrative empathy to forgotten legends. Her writing has appeared in journals of cognitive science and ethnography, alongside essays in literary magazines about the psychology of wonder. Dr. Rowan's mission: to reawaken our relationship with ancient stories as living archives of human understanding, showing us how myth and empirical knowledge together offer a deeper, more enchanted view of reality.

www.ingramcontent.com/pod-product-compliance
Lightning Source LLC
Chambersburg PA
CBHW031343230926
51848CB00002B/368